The New Digital Natives

Alexiei Dingli · Dylan Seychell

The New Digital Natives

Cutting the Chord

 Springer

Alexiei Dingli
University of Malta
Malta

Dylan Seychell
St. Martin's Institute of Higher Education
Malta

ISBN 978-3-662-50920-3   ISBN 978-3-662-46590-5   (eBook)
DOI 10.1007/978-3-662-46590-5

Springer Heidelberg New York Dordrecht London
© Springer-Verlag Berlin Heidelberg 2015
Softcover reprint of the hardcover 1st edition 2015

Printed on acid-free paper

Springer-Verlag GmbH Berlin Heidelberg is part of Springer Science+Business Media
(www.springer.com)

To God, the ultimate Coder, who created the world without using Minecraft.

To my parents, my wife Anna Maria and to our three digital natives; Ben, Jake and our latest download baby Faye.

Alexiei Dingli

To Christine, for being there in all my unconventional endeavours such as this book!

And to all digital natives of the past, the present ones and those of the future who are the ultimate agents of change!

Dylan Seychell

Acknowledgements

This long venture of exploration in such a new and ever-changing domain did not just require the sole dedication of the authors. We are thankful towards a number of people who made all this possible. A special thanks goes first of all to all of our colleagues at the University of Malta and St Martins Institute of Higher Education, particularly Vanessa Camilleri and Matthew Montebello for sharing their knowledge and experience in this area. We would like to thank Matthew Camilleri and Rachel Vella for going through our various drafts of this book to make sure that it is ultimately readable and coherent.

Furthermore, we are particularly thankful to all those who participated in requirements gathering exercises and focus groups that made the context more realistic. We would like to namely thank Christopher Bartolo, Jennifer Camilleri, Spandana Gella, Renzo Schembri, Marina Valeeva, Luke Zammit and Melanie Zammit. We naturally like to express our gratitude to all those who participated in the extensive survey that was carried out and to the others who shared their point of views on different topics discussed in this book.

Personal thanks must go to our better halves, Anna Maria and Christine for their continuous support in our academic ventures; particularly for enduring our research and writing sessions in the past years.

Contents

About the Authors

Mr Dylan Seychell is lecturer and the Head of the Department of Computer Information Systems at the St. Martin's Institute of High Education, a University of London Affiliate Centre. He is also a visiting lecturer at the University of Malta in the Faculty of Media and Knowledge Sciences. He teaches topics in Mobile Technology and Interaction Design of various levels. Mr Seychell is a co-author of various international peer-reviewed publications and a book chapter. After graduating with honours in IT from the University of Malta, he completed a Master of Science degree in the field of Intelligent Computer Systems in Mobile Technology and is currently reading for a PhD in Computer and Communication Engineering. Mr Seychell is a co-creator of the award winning project "DINOS for Smart Cities". With this idea, Mr Seychell won two prestigious international awards; European Satellite Navigation Competition 2010 - First Place in Media and CeBIT 2011 Gold Seal of e-Excellence. He also worked in various software houses and with large scale telecommunication companies.

Professor Alexiei Dingli is a Lecturer of Artificial Intelligence within the Faculty of ICT at the University of Malta. He is also one of the founder members of the ACM student chapter in Malta, of the Web Science Research group and of the Gaming group at the same University. He pursued his Ph.D. on the Semantic Web at the University of Sheffield in the UK under the supervision of Professor Yorick Wilks. While there, he worked on various large projects but his major contribution can be attributed to the Advanced Knowledge Technologies project, one of the largest Interdisciplinary Research Collaborations (IRC) funded by the Engineering and Physics Research Council (EPSRC). For this projects he created two systems which were rated World Class by a panel of international experts whose chair was Professor James Handler (one of the creators of the Semantic Web). These systems were later used as a core component of the application that won the first Semantic Web challenge (2003). His recent work in Mobile Technology & Smart Cities (2011) was also awarded a first price by the European Space Agency and the CeBIT 2011 Gold Seal of e-Excellence. He has published several posters, papers, book chapters and a book in the area. For four years, he also worked as a Senior Manager in a large government corporation where he got insight into the needs, potential and deficiencies of digital natives. During this time, he also pursued an MBA with the Grenoble Business School in France specialising on Technology Management. Professor Dingli is also the Mayor of Valletta since 2008. Valletta is the Capital City of Malta which will host the Presidency of the EU in 2017 and the European Capital of Culture in 2018.

List of Figures

List of Table

Acronyms

DTV
OTT
VOIP
VR
WWW

2DNs	Second Generation Digital Natives
AAL	Ambient Assisted Living
ACTA	Anti Counterfeiting Trade Agreement
BYOD	Bring Your Own Device
CISPA	Cyber Intelligence Sharing and Protection Act
COICA	Combating Online Infringement and Counterfeits Act
COPD	Chronic Obstructive Pulmonary Disease
DDoS	Distributed Denial-of-Service
DN	Digital Native
EC3	European Cybercrime Centre
ESC	European Society of Cardiology
GPS	Global Positioning System
HCI	Human Computer Interaction
HDTV	High-definition Television
IP	Internet Protocol
ISPs	Internet Service Providers
IT	Information Technology
M2M	Machine to Machine communication
MMORPG	Massively Multiplayer Online Role Playing Game
MOOCS	Massive Open Online Course
NFC	Near Field Communications
NSA	National Security Agency
OS	Operating System
PC	Personal Computer
PIPA	Protect IP Act
RDSR	Relative Direction based Sensor Routing
RFID	Radio Frequency IDentification
RTS	Real-Time Strategy
SDK	Software Developers Kit
SL	Second Life
SOPA	Stop Online Piracy Act

TBS   Turn-Based Strategy
UCD   User Centered Design
VOIP   Voice Over IP
VR   Virtual Reality
WWW   World Wide Web

Chapter 1
Introduction

The first generation of Digital Natives (DNs) is now growing up. However, these digital natives were rather late starters, since their exposure to computers started when they could master the mouse. In the early years of the new millennium, the penetration of computer technology was low in most homes [58] [64]. Laptop computers were slowly becoming mainstream but still, they were strictly secluded to the workplace. The main mobile device in existence was the mobile phone but still, it was just a glorified wireless telephone [265]. Smart phones only penetrated mainstream use towards the end of the first decade [339] and popular phones such as the Blackberry were mainly restricted to corporate use. In those days, broadband was still a luxury [144] and as a consequence, the rich hypermedia sites with videos and complex animations (which we are accustomed to today) were far from being conceived due to the limitations imposed by the technology of the time. Towards the end of that decade, the standards conceived by the W3C started to gain mass adoption and websites were slowly turning into web-apps. It was the transitional period when the web was morphing into what we refer to as being Web 2.0. This web in not a new network of computers but rather a fuller realisation of the existent infrastructure. At the time according to [261], social media was still in its infancy and the social web we have today did not even exist. So really and truly, the first generation of DNs were mildly introduced to technology. They could really start using a computer when they learnt to handle a mouse or when they started to read. Even when they started using a PC, the content available for them was next to nothing. Computers where still locked to a particular Operating Systems (OSs) so the software available was extremely limited. Laptops, being considered as a business machine at the time, where generally out of bounds. The web, was used as a reference tool and the always connected paradigm was still in its infancy. The fact that wifi was very limited meant that the machines connected to the internet were much less than we have today. Thus, children could not gain access to online content (which at the time was rather limited) easily. Mobile phones did not really have an exciting colourful interface and the only way in which they could be used was through the intricate combination of their buttons (Touch displays gained popularity towards the last quarter of the first decade when the first generation iPhone was released).

© Springer-Verlag Berlin Heidelberg 2015
A. Dingli and D. Seychell, *The New Digital Natives*,
DOI: 10.1007/978-3-662-46590-5_1

Thus, the only use which the first DNs had of mobile technology was as an expensive teether. The parents of these DNs were probably born around three decades before when technology was still very archaic. Just consider that most of them did not even have access to a computer when they were kids and they were introduced to the Internet when they were in their late 20s. So the support which they could offer their kids was rather limited.

The second generation of Digital Natives (2DNs) who were born around the end of the first decade can be considered as a new breed of digital citizens. First of all, they have an unprecedented access to technology [285]. A personal computer (normally a laptop), can be found in most homes and adults normally own more than one personal digital device (these include smart phones, tablets, etc as can be seen in [123]). A substantial amount of homes in the western world have broadband which is provided via wifi. This means that users have access to high quality content on any device. Further still, the location of the device is not important since wifi provides the freedom to consume digital content from anywhere. Finally, the fact that a lot of devices these days come with a touchscreen means that the basic requirement to use such a device is a finger. This leads us to the scenario where one year olds manage to master the intuitive touch interfaces of their tablets whilst sitting comfortably in their baby bouncers. They do not even need to grasp the basic concepts of a language (As it was the case with the first generation DNs) before operating a device. The fact that they can touch an interface and the interface is responding back to them is enough to stimulate them and encourage them to explore further (even though the communication with the device might be meaningless in the beginning). The parents of these 2DNs also experienced the introduction of the personal computers in their homes. They not only remember the original versions of space invaders or pong at the arcades but they were lucky enough to own a personal computer where they could play them. This creates a new openness towards technology and a drive towards owning new technological devices. Thus, they tend to encourage their children to use these devices from a tender age, they play with them digital games and also support them when they face difficulties. This is made even easier with the availability of digital content which can be downloaded from the web. The Internet provides access to a myriad of educational and entertaining content which is free, thus keeping the cost of usage low. Further-still, the push towards ubiquitous computing and controller-less interfaces (such as the Kinect[1], Leap Motion[2], etc) allows these children to interact with a machine in a way which was unconceivable below.

Because of this, we will start by first presenting these digital natives to the reader. We will delve into their psychology and what makes them different from other generations. This will help us understand and appreciate why DNs act in the way they do, why they take certain decisions and why they have different expectations. We have to understand the context in which they were raised, the instant society [267] whereby anything they require is within their reach. They are not mesmerised by

---

[1] http://www.xbox.com/en-US/kinect
[2] https://www.leapmotion.com

a handful of TV channels because now they have thousands of online channels to chose from. Their relationship with technology is different from that of past generations because for them, technology is not just a tool but rather an important extension of their life according to [342]. It is used to communicate, to learn, to express oneself, etc. They are not afraid to use it [205]; as they are not afraid to expose themselves and their views on these online fora.

We cannot understand the digital natives without also understand the context in which they live. In particular, we will examine the paradigm shift which occurred throughout the years and how this effected the relationship between men and machine. Our journey will take us from the microprocessor, batch computing, time sharing, networking and will keep on going until we cut the digital chord thanks to wifi technologies. Our 2DNs are no longer stuck to the machine via a wire. This has obviously effected the way in which we look at our world.

How can we distinguish the office space from the home space?

How can we distinguish between office time and home time?

The answer is simple, these distinctions are becoming extremely blurred as can be seen in [75] and it is becoming very hard to distinguish between them. Whilst at the beach, I can still check my emails and chat with a dozen friends. So our lives are quickly changing and the lives of our 2DNs is very different from the one we used to live when we were still young. Space and time are not the only two dimensions that have been invaded and distorted. Just think about the physical and the virtual. Most of our documents already made the leap. They are stored in files which we send by email or through instant messaging. But before, these documents had a physical location inside a storage device (the most common of which being a hard drive). This device stored files in folders, thus replicating the metaphors which we normally find in an office setting. In the past years, the notion of a physical storage device is quickly disappearing. We have all been to the funeral of floppy drives and DVDs. Now the time has come for the hard drive, which is being replaced by the cloud [306] which is another metaphor of a virtual storage, far far away. Notwithstanding the fact that we don't see it, we still put our most valuable data on it and pose our trust in the company that provides us with cloud services. This is a huge leap of faith and only time will tell if we were wise enough to adopt such a system. Same thing is happening to the notion of a personal computer. We're moving towards a personal device whereby different people own various devices and use them to access their data stored on the cloud. Once again, we have to cope with different devices, having different capabilities, whose use is gradually changing. Before they were the source of power, today they are just a window over a remote machine somewhere over the internet. Our precious data is located in that remote machine, in a place which is unknown to us, but on the other hand, it is safe from hardware failures and other vulnerabilities. This poses new issues related to the trust we pose onto the service provider and could open a myriad of sensitive issues if that trust is breached.

It is obvious from what we have seen so far, that the world is turning into a complex place. Our existence is shared between different realities; the physical world, the online world and maybe hybrid worlds. In reality, it is more of a set of different realities coexisting together and to understand them better, we will look into blended realities. First of all we'll examine the different devices available; not just personal computers (PCs) or laptops but also tables and all sorts of wearable devices. We have seen in recent years the rise of the smart phone as a contender to the role of the PC. In fact, the processor of these devices have become so powerful that some UBUNTU phones[3] even dub as a fully fledge PC. But the future seems to be moving towards the creation of more intimate devices; such as digital jewellery or even digital glasses. The trend is more towards shrinking technologies probably to such an extent where technology will disappear to the naked eye. This technology will become embedded in our daily lives; in our clothes, in our homes, in our vehicles, etc. Essentially, we are ushering the age of ubiquitous computing. This would not have been possible without the recent developments in nano and micro electronics, without the advancements in wireless technologies, without web 2.0 protocols and without the rise of cloud computing.

These advances have an obvious effect on the applications we use. Productivity apps strive towards automating our lives by providing us with real life but context dependent information. To further enhance these apps, we are augmenting social information with existing data thus creating a powerful mashup which was unconceivable before. A simple map can be easily augmented with real time traffic information together with comments from our peers. The possibilities offered by such a system are practically endless and provides new services aimed at making our life better and easier.

But to nurture our 2DNs in this digital world we need a whole new approach. That is why we will focus on their day to day life starting from the first instances of their existence. We cannot expect these individuals to be treated as the first generation of DNs because they are very different. They have an unprecedented openness towards technology as can be seen in [309] because they were immersed in it from when they were infants. Even though it is hard for our brains to multitask, these 2DNs seem more capable of coping with different streams of multimedia information. It is not the first time we see them reading whilst listening to music and watching TV. So when we place these 2DNs in the classroom, we should not be surprised if they get bored [30]. Traditional tuition is monotone and monochrome (limited to writing on a board). It was only in recent years that we've seen the mass introduction of technology in the class room such as the installation of the Interactive White Boards, Projectors and the more recent introduction of tablets. However, now that we have technology in an educational setting, the big question is how to use it effectively. It is no secret that some educators feel at lost when presented with these technological innovations. Even young educators, fresh from University, need some time to accustom themselves to these new pedagogies. Given that we now have the tools, the past years, have seen a general effort towards creating more effective teaching methodologies. In particular, we have seen the rise of gamification in eduction which is the idea of utilising game mechanics to teach. In their essence, games are entertaining and manage to capture the attention of the players. Educators want to do the same thing, they want to pass on their knowledge to their students through a fun setting

---

[3] http://www.ubuntu.com/phone/ubuntu-for-android

which captures the attention of the individual. Further-still, gamification allows educators to also offer a personalised approach. Players progress through a game at their own pace and in a similar way, we want students to move on through the educational system by stretching the limits of their own abilities. Ideally, when these 2DNs leave our educational system, they would have developed into young adult whose talents have been developed to the full.

When reaching adulthood, 2DNs start looking for a job and the workplace too has to change [343] in order to offer them the best environment, where they can increase their productivity. It must leverage on their technological skills and align their output towards corporate targets. If management does not manage to do so, but expects workers to adapt to the company's culture, it will be only wasting a lot of throughput. One has to keep in mind that in our knowledge society, the most valuable asset is the employee. Once an employee walks away, it takes with him precious information and finding an adequate replacement is not that easy. That is why various companies try to externalise the knowledge within the employees and code it up in an electronic system. But this is not always possible, both because of the nature of the knowledge and also because it can be financial prohibitive. Also, if they manage to do so, they are effectively making the employee a redundant fixture in the operations of the company which might reduce the job security of the individual involved and as a consequence, effect its output. Because of this, the best strategy is to change the organisation and make it reach the expectations of these 2DNs.

Once they start earning a living, they also start spending. One has to keep in mind that these youngsters have been raised with hundreds of TV channels and most of them got used to on-demand high quality content. They are connoisseurs in different forms of entertainment, so the type offered by different venues has to meet their expectations. They don't want to be spectators but rather protagonists, who delve further into the meaning behind the artefacts in order to reveal the concealed story behind them. In reality, visiting an exhibition or attending a show, is no longer about the performance or the artefacts on show, but rather about the experience conveyed to the audience.

Their demanding lifestyle is not just felt when they go out but also in their day-to-day lives. Their house is slowly morphing into a smart home, with gadgets hidden in every corner in order to control and help them have a comfortable home. The smart home will have technological aides in the kitchen, bedroom, bathroom and all the other rooms of the house. Sensors will be strategically located in order to ensure that their homes are energy efficient (thus saving them money) whilst offering them higher comfort than it was possible before. If we just think about the heating or cooling within the house, it will adjust itself to the number of people in specific rooms, it will monitor the lifestyle of its owners thus switching off when no-one is around. At night it will ensure that the person is comfortable and reduce it to an absolute minimum only to switch it back up in the morning just before the person wakes up. Personalised information will be downloaded on a daily basis from various news sources and an ad-hoc newspapers will be collated made up of the most interesting news items personalised for the user. Appliances such as coffee machines or washing machines will work when their owner is resting or while he is away.

This will ensure that food or washed clothes will be ready just in time. The benefits to the user, will keep on increasing with time and the smart home will easily change into an ambient assisted living space. Rather than just monitoring the user and make his life comfortable, the system will also take care of his wellbeing and assist him in his daily chores. It will instruct the user to take specific pills at particular times. It will monitor the temperature, pressure or glucose level. If the person faints, the system will detect that unusual state and alert his next of kin or the nearest hospital. From being a passive aide in the house, the system will turn into an active assistant responsible for the wellbeing of the individual.

Apart from their physical lives, 2DNs have a very active digital live. First and foremost, they are online citizens and as such, they have different rights and obligations. Most of them do not feel part of a particular country but their life is based around this global cyber village. They can access information from all over the globe and send information to any part of the online world within seconds. Their time online is as important (if not more important) than their life in the real world. Keep in mind that cyberspace is the place where they interact and socialise with their peers. Because of this, governments are very much concerned about these virtual boundaries between countries. In most cases, they can't be defined clearly especially when a website or web-service is provided through multiple servers geographically located in different countries around the globe. That is the reason why different countries are pushing towards digital legislations. In the past, we've seen numerous attempts at granting greater control to governments and also various clamorous failures. Only time will tell us the extent which government agencies manage to monitor and control our lives. But digital legislations have their positive aspects as well. They can give greater rights and empowerment to citizens whilst also making them participate in the legislative process. In fact, there are a number of countries which allow people online to post suggestions and amendments to bills via online means.

With people migrating online, we're also seeing other groups migrating such as politicians and criminals. The rise of social media in the past decade has seen a growing number of elections being fought online. This trend has been spearheaded by the American Presidential elections whereby millions of people were being approached through online means. These advances ensured that people felt for the very first time a connection with their politician. In large countries such as the United States, it is unthinkable or extremely difficult to meet a presidential candidate. It is even more difficult to meet him or her after the election. Notwithstanding this, social media creates a direct virtual channel with the political candidate. It works two ways, the voter can query the candidate and the candidate can make requests to the voter (such as the case of the last US elections whereby millions of dollars were collected via crowdfunding initiatives).

The underworld of crime too migrated online. In fact, cyber attacks are very common and they're becoming harder to counterattack. The level of sophistication reached by these criminals is extremely high and governments world wide together with citizens are going to great lengths to protect their precious information. This information can include anything from financial information (such as bank accounts) to personal information. The consequences of these crimes are unimaginable. By leaking a list of passwords or bank accounts, millions of dollars might be in jeopardy which can easily lead to the ruin of large institutions. But apart from the financial implications, the risk of identity theft is also growing whereby the online identity of a person can be stolen and used to commit other crimes. There is also the risk that online criminals move over to physical crime. We've seen various incidents around the world where people have been kidnapped or even killed as a consequence of their online activity.

But even though we have these threats lurking over our head, the future does looks bright and exciting. Because of this, we need to investigate the shift between the 2 generations of digital natives. We will analyse the way in which these two generations differ from each other and we'll study how the world will change in response to the new digital natives. Technologies such as the Cloud, Artificial Intelligence, Mobile Devices, Quantum Computing etc will all play a massive role in our future lives. That is why we need to engineer the future for these 2DNs. We need to understand their requirements, their needs, their expectations and aspirations. We need to work on it in order to create a better future for everyone. But in particular, we need to help them develop their full potential in order to reap the maximum benefits from their work. After all, the future is theirs; to shape and mould this new digital world.

Chapter 2
Who Are the Digital Natives?

2.1 Introduction

We are increasingly realising that the exposure of children and young people to technology is strongly affecting the way in which society develops. Individuals, who do not find the complexity of the digital era and constant updates in the field of technology problematic, are generally referred to as 'Digital Natives (DNs). This notion was introduced by Marc Prensky [254] when he defined the gap in the way these two generations deal with computers and the Internet. Prensky's essays [256] [254] discuss the concept of Digital Natives and Immigrants in the education context but this book takes it even further.

Digital natives are today's young people who were born into the digital era and are growing up exposed to the continuous flow of digital information. Digital natives are a generation or population growing up in the environment surrounded by digital technologies and for whom computers and the Internet are natural components of their lives. They do not need to familiarise themselves with the technology by comparing it to something else. On the contrary, they propose new ways of thinking about how technology can be effectively used. Digital Natives perceive the world through different eyes: what is a novelty for digital immigrants, is something ordinary for digital natives and ultimately an integral part of their lives. Nevertheless, there are individuals, who fall within a grey area since although they were not born in a digital environment, they still manage to integrate in this digital environment. These individuals are also affecting the way organisations operate today due to their ease of use of technology at hand.

This study aims to investigate the characteristics of the digital natives while aiming to elicit clear distinctions between individuals belonging to different populations of use. Each group brings with it a new set of skills that can be cleverly used to optimise present processes. However, it might also be the case that certain skills might be lost between generations. In order to discover these distinctions this study also studies the trends that these generations engage themselves in.

© Springer-Verlag Berlin Heidelberg 2015
A. Dingli and D. Seychell, *The New Digital Natives*,
DOI: 10.1007/978-3-662-46590-5_2

2.2   Psychology of Digital Natives

The key issue surrounding digital natives is whether or not they are a population or a generation [236]. In any case, the focus of this book is about the 2DNs, children born in the wireless age, in other words, in the start of the twenty-first century. Since these individual are young and newer groups of digital natives will always be young, this section is dedicated to the link between Developmental Psychology and the study of Digital Natives.

Vygotsky [318] introduced the idea of individuals developing as a self through the influence of others. This theory of social development states that social inter-actions help children develop. Vygotsky explained that children do in fact develop through play and experiments particularly when these are carried out in the company of others. This is very relevant in the age of devices where besides interaction with others, there is also significant interaction with devices. There exist applications that provide children with different forms of exercises and reaffirm their potential when they get a correct answer. Children also need to carry out tasks independently and enquire when needed. However, with the development of technology, they are now being supported by devices which are capable of tailoring their approach according to their learning style. Our studies [85] also found that 63% of children between the age of 7 and 12 are mobile phone users with 31% of these being tablet owners. The study also showed that in other cases, children make use of tablets that are owned by adults. Vygotsky's idea of proximal development [318] also illustrates that there are certain tasks that children can do with some help and others without help. Our study also shows that in 55% of the cases, children under 12 do not only use tech-nology without asking for help but the adults seek their assistance when they are using digital devices.

One of the leading authors in the field of developmental psychology is Piaget who explored the idea of stage based cognitive development. In his work, Piaget [247] linked age with the way the human mind develops. He also explained that the rate of development depends on the abilities and inabilities of the subject in question.

Table 2.1 shows the developmental stages that take place during childhood. Piaget's theories also underlined the principle that children can carry out the tasks when they are ultimately mature enough to do. There is a strong link between major observations explained in this book in relation to Piaget's stages. There is a link to how toddlers manage to swipe along screens full of icons, choose their favourite application, use it and when they are done, they press the home button to go back to the menu and choose a new application. This falls in line with Piaget's observations of the sensorimotor stage where children first experiment by trial and error and then they become goal oriented (choosing an application) and apply symbolic thought processes (icon to load an application and pressing the home screen to go back to safety).

Similarly, during the pre-operational stage, children exposed to technology would start to strongly familiarise themselves with technology. During this stage, their mind is on the fast track capable of learning new languages where they develop a vocabulary of around 800 to 900 new words per year [27]. This is often equated

Table 2.1 Cognitive Development (Adapted from [213])

Stage	Period	Characteristics	Notes
Sensorimotor	0-2 years	Symbolic Thought; Goal Directed; Object Permanence	Trial and Error
Pre-Operational	2-7 years	Develop Speech; Able to pretend play; Egocentric	Develops Intuitive though; Dominated by perception
Concrete Operational	7-11 years	Conservation; Class Inclusion	Able to think logically
Formal Operational	11 years +	Able to think logically and abstractly	Idealism and able to solve hypothetical questions

to their ability to absorb new features in technology. Language does in fact share significance with the learning of technology particularly because language is in itself a process of free creation [59]. This notion of free creation in language comes along through the way children use grammar which they were never taught and therefore language is learnt just like the use of other organs is learnt. This is then reaffirmed when adults approve through different gestures and sounds any form of positive progress children do when experimenting with new grammatical constructs [292]. This is one of the occasions where Piaget's stages and Vygotsky's ideas meet particularly because staged development is being reinforced by immediate social interaction. In the pre-operational phase, children also start to demonstrate their capability of organising different objects logically [21]. This is particularly visible through the different games which children play while demonstrating an evolving logical ability. In another dimension of this stage, children also start to develop their egocentric self by becoming more self-aware of themselves and their capabilities without being able to see themselves from the perspective of others. Therefore, if their immediate parents or guardians reaffirm their ability to grasp new features in technology, children are then practically transformed to 'aware' digital natives. This is backed by the result of the studies that were covered during the writing of this book that show a significant 55% of parents stating that their children under the age of 12 are more technological knowledgable than themselves and confirm this by asking the latter for help in this matter. If this is clearly transmitted to children and reaffirmed, their developing egocentric self would then absorb this detail and consolidate the fact that they are native speakers of technology.

In the subsequent stage of Concrete Operation, children start to apply logic to the tasks that they would have been carrying out till then, linking them carefully to the notion of language that is being developed in parallel. At this point, children are also able to logically decide upon certain choices presented to them such as for example their ability to understand that quantities of states such as liquids can take different forms or shapes [21]. However, this ability is limited to only concrete objects that children can actually see or perceive. On the other hand, in the fourth stage of formal operational, children can start to think both logically and abstractly; therefore starting to manipulate their own ideas without clinging to concrete objects. In these last two stages, children are less egocentric and would start to listen to what other people are saying about them.

2.3 Terminology

This is a relatively new and evolving field of study and to date, there is no universal agreement among researchers about the terminology used. This section aims to define the key terms used in literature and this book.

This terminology was traditionally bound to the decade of birth of an individual but on the other hand not all researchers agree on the decade that defines the digital natives thus leading to the loose classification via generations. Such classifications are mostly based on rhetoric and it is therefore difficult to harmonise the said terminology. With the increase in the use of this terminology in various spheres, from education to politics, interest from the research field has also increased in this topic. This lead to the emergence of more empirical research that delved deeper into the classification of people vis-a-vis the use of technology.

We therefore classify the various research efforts into 2 categories: Visionary and Empirical. Both are very useful in the discovery of technology usage and in fact Section 2.4 is dedicated to the analysis of such approaches.

Digital Natives basically refers to those individuals who were surrounded by technology from their birth. Being surrounded means that they were in reality exposed from their early days of technology. In this case, age is only important because these young people were simply born when technology was available. However, this does not mean that all born in the same period were also surrounded by technology. Such reasons would therefore shed light on the relationship between the Digital Natives and technology in terms of immersion. These people think and reason in the context of technology and with basic technological assumptions. Besides reference to Digital Natives, one may also find reference to another group of people that is referred to as the 'Digital Immigrants. This is Prenskys term for people who were not born in a world surrounded by new technologies but they were only exposed to it at a much later stage in their life. Digital immigrants did not grow up playing computer games, downloading movies or sharing information online. They had to learn about different technologies and its use when they became adults so it took them additional time and effort to adjust to the world of technology.

Since digital natives and digital immigrants belong to two distinct usage groups, there is a tangible gap between them, which consequently results in specific distinctions. The main distinction is in the way these groups engage with new technologies. Prensky [256] draws a comparison between natives and immigrants: Some immigrants accept that they are not knowledgable about this new world and use their children to help them learn and integrate. Other (not-so-flexible) immigrants spend most of their time grousing how good things were in the old country.

These two particular terms have been introduced to highlight the importance of technology in the lives of these two groups and now we are starting to witness the effect that the members of these two generations are leaving on our society and organisations. This books even takes this terminology to the next level when we introduce the notion of the 2DNs. This notion is introduced in more detail later in this chapter.

All arguments surrounding the notion of Digital Natives can be ultimately explained in terms of space and time. These may therefore be classified with respect to when people were born and also when people were born i.e. a place which provided them access to technology, thus for them, technology usage was a ingrained habit. When dealing with Digital Natives, technology is also often equated to language in terms of immersion. Prensky said that Digital Natives are in fact native speakers of technology and also went into the depth of saying that Digital Immigrants speak an accent of that technology.

Technology and Language do however share particular features that make such comparison plausible. We learn a language just because we are surrounded by people who speak that language. Similarly, if we are surrounded by technological devices from day one, these would form part of our daily environment and just as we learn to interact with anything as simple as a cup of water, we learn how to interact with technological devices that might be as easy to use. Children expand their vocabulary pretty rapidly [146] and they can similarly absorb significant information that is related to technology. On the other hand, just as in languages, learning how to use technology does not really mean that such children can handle technology in different context and applications. Simply because a child learns and remembers a word within a context, the child would not be really capable of understanding all the usages of that word. In practice, if a child learns a word, it would be difficult for him/her to learn any metaphors related to that word or the political correctness behind using the designated word. This therefore brings into the scene the notion of skills vis-a-vis technology. Just because a child or a Digital Native can use a technological device, it does not mean that the child can apply the device to a more efficient context in the real word.

2.3.1 The Digital Natives

The term digital native is a popular notion that describes the reality we are living in nowadays, and it is very important to know what is actually hidden behind this terminology.

Prensky sets a clear scenario of individuals being born right into a world that just adopted computers and the web. He defines Digital Natives as those who are born in this reality and are native speakers of this technology. Many [332] [236] agree with Prensky that as a result of this sole exposure to technology, Digital Natives do in fact think differently. Digital Natives think in random access like hypertext, use instant information for judgement and so on [254]. This shows that the way Digital Natives think is directly influenced by the technology they are using.

Many researchers also agree that the beginning of the digital era set off this new generation, due to the great innovations that were presented to the world at the early 1980s [236]. This was indeed a big leap for technology, when the world saw the dissemination of personal computers, the improved utilisation of the Internet and later on the World Wide Web in 1991 [22]. During the year 2004, the world witnessed the evolution of the web towards what is now known as the Web 2.0, and as discussed in Chapter 4 the escalating use of mobile devices that evolved into touch devices in the same decade. Later on in 2008, the music industry was controlled by the iTunes store in the United States [236] and it paved the way for a change in mentality with regards to the purchase of musical records.

Just after reading the previous paragraph, one can quickly conclude that other technological milestones were left out. It is important at this stage to emphasise that technological developments do not wait for us to properly classify their users. It is therefore a technological driven scenario whereas a technology that catches pace, hooks in new users. The lifestyle of these new users changes accordingly and it is at that point where one can or may start to classify user groups.

This sheds light on the definition of Digital Natives. Which technology defines digital natives? One may argue that the first commercially available portable digital device was a calculator in the 1970s. Does this mean that those who were born in the 1970s and made use of this device are digital natives? Others may argue that it is after all related to the technological environment and a set of few devices does not represent the said environment.

Some researchers suggest that digital natives are 'people who were born during this period, between 1980 and 1994 [20], were labelled as "digital natives. This generation was given other similar terms like the 'Net Generation [307] and Millennials [149], which were also describing those who include technology through their daily life. Looking back at the technological developments that are explained in Chapter 4 we also propose that this bracket is further increased to cover individuals who were born till the year 2000. In Section 2.4 we also demonstrate that this is not simply related to the age of birth but also to other factors such as exposure to technology.

2.3.2 The Digital Immigrants

With the notion that Digital Natives are the native speakers of technology, one naturally asks the question: "What about those who are not native speakers?". These are the individuals who simply were not born during the digital era but still make use of technology. One can also zoom in further into such a claim. Prensky [254] simply defines Digital Immigrants as those who were not born in a 'digital world' but adopted a significant number of technological aspects. This could be due to various reasons including the lack of technological knowledge, the inability of learning how to use such innovative devices and not finding any particular need to make the crossover from previous methods.

In Prensky's metaphor, the Digital Immigrants are those who although are not native speakers of the technological language, they still have to 'live' in the land of technology where Digital Natives are the native speakers of this language. Digital Immigrants make use of traditional methods to develop and represent their thinking. Plainly said, Digital Immigrants are those who do seek the friction of a pencil onto a paper when sketching their ideas. Yet, they also need to at various stages represent their ideas via technological media. In practice, this relates to a manager who needs to explain certain goals to his team. We choose to organise this scenario into three parts: Planning, Design and Delivery. A Digital Immigrant would normally plan his presentation and idea on a piece of paper, even if it is the backside of an envelope. Subsequently, when the concept is clear enough, this manager would switch to a presentation software to illustrate his idea in a clearer and more attractive manner. During the actual delivery of the presentation, there are two different approaches a Digital Immigrant might take. The first approach is the 'back to paper' approach with the manager following notes written down on a paper while having the actual presentation on a screen behind him. The second approach is the 'digital approach' where using a mobile device, the manager would be able to switch between slides while reading notes on the mobile device. This sheds light on the classification of Digital Natives and Immigrants. There are points were immigrants are comfortable with traditional media, yet at some point, they consciously and comfortably resort to modern digital methods.

Since the digital natives have statistically not taken over yet, the methods currently employed by Digital Immigrants are still relevant and widely used in various aspects of life. Prensky chose the 'Immigrant' terminology carefully. As a matter of fact, Digital Immigrants share a lot of features with other immigrants who move to different countries. Different countries have different policies, cultures and above all, house other individuals who are native in that environment. The same applies to the digital metaphor. Although Prensky focused on the educational domain, this reality is also propagated across other domains.

If we analyse the context of the Business world, there are two types of organisations. There are those that are 'Born Digital' and others that were not but are in the phase of 'Becoming Digital' [303]. A subset of these organisations will simply keep growing and are made up of executives who are Digital Natives. In this case, the company is firstly born digital and besides, we add that it is organically digital in all of its processes. Stein and Lipsher [303] discuss the scenario where the executives are both Digital Natives and Digital Immigrants. They argue that the companies that are born digital have a relative advantage over the other companies for two main reasons. The first reason is that today's cost of starting a digital company is only a fraction of what used to be a decade ago. Secondly the companies that are born digital, have also the advantage of a free mind set in the modus operandi. This shows that there is a shift towards the importance of human capital in modern organisations [303].

At this point, one can imagine the scenario of two companies. One that was set up in the early 1990s and a recently established digital company that is composed of digital natives. Generally speaking, in terms of age, the first company was set up when the executives of the second company were in their early childhood. This is an interesting scenario where these two companies are in competition and in other words, it is also a competition between Digital Natives and Digital Immigrants. Stein and Lipsher argue that companies that are in the process of becoming digital, are employing individuals in the role of Chief Digital Officer who would be responsible for this transition [303]. This is also an opportunity to inject new blood obtained from the digital natives into companies that were not born digital. There are also different strategies which companies can take in this scenario and context yet it is beyond the scope of this book to explore such methods.

2.4 Classification

The previous sections of this chapter explored the terminology in this research field. Terminology is easily transmitted and explained by the use of metaphors. However, classification proves to be more challenging. Digital Natives and Immigrants were primitively classified by their year or decade of birth. Experience showed nonetheless that there are other factors that have to be considered.

This section aims to expose the different interpretations of the same terms while showing a harmonised set of terms that are used in this book. The challenge with this classification process is that the science related to it in the first place is still in its infancy. The concept of an electrical computer has only been with us for around 70 years and technology was only widely available and affordable in the recent 2 or 3 decades. As explained in more detail in Chapter 3, we are now experiencing something similar to what was experienced in the Newtonian days of the the rise of rationalisation and physics.

In this exploratory stage of the digital era, we are motivated to get to know better who uses the computer and what they have been developing in these last decades with such a significant impact on the world. In the days when the computer users were only those people who built the machines, researchers were understandably not really concerned about the mass users. Their profile was generally the same and they were not a significant cohort in the social sphere of those days. With computers becoming more and more ubiquitous, the social fabric of society started to be influenced and affected by this new reality that could no longer be ignored.

Earlier in this chapter, we referred to the research in this field as visionary and empirical. In the scenario where no literature explored the reality of Digital Natives, it is understandable that the first authors, such as Prensky [254], who explored this idea had a visionary approach. In retrospect, this work served as a motivational ground that set the foundations for discussion about this topic. Once the topic gained momentum, one started to see empirical research that started to enquire these visionary claims and survey the actual reality.

Most of the visionary arguments were already explored in previous sections. These basically contain conflicting arguments about the age bracket in which DNs fall. According to Marc Prensky, Digital Natives are native speakers of the digital language of computers, video games and the Internet [254]. In the same article, Prensky suggests that age is the defining factor in defining the level of digital nativeness. However, Tapscott [307] proposes that this should be defined by the exposure to technology rather than by age.

Helsper and Enyon [139] carried out an empirical research that sheds light on the visionary arguments put forward by Prensky and others. This effort proposed that the classification of Digital Natives should be based on 3 metrics: Age, Experience and Breadth of Use [139]. Age is the key metric that was used in the visionary literature. However, while this indicates a course for navigation, it leaves the sails fairly loose and prone for tangling. The other 2 metrics are then used to asses the degree of nativeness and technically highlight the difference with immigrants. Experience considers the situation where irrespective of the age, an individual would be exposed to technology for a long while and thus confident with its application and use. As a further focus of the first 2 metrics, Breadth of Use measures the integration of technology in the everyday life of users irrespective of their age and experience [139].

This empirical study provided new tools through which one can look at this field of study since it untangled the classification issues that were till now bound to the age of the user. When linking these findings to the time and space reality of digital natives, one would then reaffirm that digital natives are in fact the young users who were exposed to technology and are therefore comfortable with the technology that they are surrounded with and thus using.

2.5 Digital Methodology

Differentiation of Digital Natives and Digital Immigrants might not be evident across all aspects. As a matter of fact, the boundary between the two is at times blurry and it is therefore challenging to classify an individual when there is not enough evidence available. This section aims to illustrate the key aspects that help in the classification of individuals. With the emergence of new technological devices and their interconnectivity through the realisation of ubiquitous computing, there are today different ways of carrying out the same task. Modality can take digital form or the traditional form. Nevertheless, when in the digital form, modality may also change in accordance to the devices available. A practical example would be planning a trip on a map. In the traditional form, a physical paper map would be used to familiarise oneself to the area and plan the trip upon it. Later, one would take the same physical map on the trip and use it to go around. However, if one had to assume a totally digital form, things would be pretty different. At planning stage, one may plan the trip on an online map located on a desktop computer, having all information at hand from the Internet. The user would have used his credentials to log in this online map application which would enable him/her to save progress. This then automatically renders the map available on the tablet and mobile phone for the user. During the trip, the user would not carry the desktop computer that was originally used to plan the trip. When on a train, final preparations might take place on the map application which is located on the tablet device because it has a larger screen. When the user arrives at his destination, he would then walk around the city without the need to consult the map. However, if the tablet is not available and the user gets lost, he may still access the same map on the smart-phone and find the way back using the map together with other sensors such as the GPS and the compass. This flexibility in digital modality would enable the user who is familiar with a computer to have the information available and reason or plan the problem using different devices.

2.5.1 Social Identity

Identity was once something very straightforward: people expressed themselves mainly through their personal characteristics, appearance, clothes, hobbies and interests. On the other hand, individuals who spend most of their time in cyberspace do it in a similar way by expressing themselves in the real world whilst also extending and complementing their offline social life by making use of the digital environment. In other words, an identity for a digital native is a synthesis of real and online expressions of oneself. People might express their personal characteristics online in more ways but in its core, their personal identities will not differ much from the one they have in reality.

The way of expressing oneself is the main difference between digital natives and immigrants. If digital natives are able to shape their social identities not only in real space by changing the colour of their hair or clothes but also through online applications or social networks like Facebook or MySpace. The digital age allows digital natives to change many aspects of their personal identities much quicker and easier than it was before.

It is important to understand that the world wide web does not change the idea of the identity altogether, it simply provides possibilities to create an identity online, for example, a new profile in a social network, where digital natives could present themselves in a way that could be striking different from the way they present or express themselves in real space.

Digital natives tend to change aspects of their personal and social identities almost constantly [237]. Digital natives can experiment with their identities; they can create multiple identities online. Natives update their avatars and profile pictures as frequently as they change their clothes or hairstyle. Natives constantly add and sometime delete friends. Making friends became so much easier in digital age. If digital immigrants still prefer to make friends in person and more likely within their current location, digital natives are open for friendship with people from all over the world.

Communication has become easier in many ways, the distinction between online and offline is less visible. Digital natives express themselves through their posts in social networks, blogs, videos and music which they share or upload on YouTube. Other natives change some of the aspects of the digital identity: friends can post items on the Internet and associate the names of their friends with them. Identity formation among digital natives is quite different from identity formation among digital immigrants in the sense that digital natives are ready to experiment and reinvent their identities. Through public spaces on social networks, digital natives appear more open-minded, they learn how to socialise and express themselves, experiment with, develop, and learn to evolve their talents and present their identities.

The nature of the identity is changing in the 21st century. It affects not only digital natives who already do not distinguish between online and offline identities, but also digital immigrants who try to acquire the digital literacy skills to keep abreast with young people.

2.5.2 Travelling

Mobile technology is widely used in tourism and it has been 'one obvious application area' of mobile technology [35]. In our previous work [82] [80] we have shown how the mobile nature of tourists and the respective nature of travelling renders mobile devices into an invaluable tool. The problems faced by tourists are related to what they should do and where to carry out an activity. Furthermore, tourists also need to know how to go around the foreign environment [35]. The availability of information on mobile devices coupled with their capability of integrating sensors with software intelligence solves problems that tourists face while travelling [82]. This is coupled with the ever improving computational power of mobile devices which boosts the creation of more mobile applications thus offering a better experience to tourists [68].

With the increasing penetration of mobile devices, further investment in mobile applications also strengthened the link between the technology and travellers. Traditionally, the paper map and a photographic camera characterised an average tourist. With maps being replaced with intelligent and up-to-date mobile applications that know the user's location together with navigational assistance, tourists are finding it easier to explore unknown spaces by simply making use of their mobile devices. Similarly, smart-phones have integrated high-resolution cameras that make it easier for the average tourist to capture any moment, instantly. Furthermore, mobile devices add the further advantage of internet connectivity thus enabling the user to instantly upload and share the photo that was just taken. Sharing capability is also available in navigational applications where users can share the paths they took while travelling in a particular location.

The nature of travelling with such emerging technologies is rapidly transforming itself by particularly allowing visitors to have more reliable information, make more informed decisions and also carry less items with them.

2.5.3 Communicating

The main motivation behind the most recent developments in technology was and is the need of communicating with others. The web itself and its availability on mobile devices highlighted the reality that is discussed in this book. The emergence of web 2.0 enabled web users to upload content and thus share their experiences. Connectivity has become synonymous with computing.

In the focus groups that were carried out with the second generation of digital natives, particularly individuals who were born in the 21st century, we asked the question 'what is a computer?'. These sessions were held in laboratories with numerous desktop computers. When asked this question, they all swiftly indicated the machines in front of them. We then presented the participants with two images, one of a desktop computer and the other of smart-phone. They practically all agreed that both were computers. None of the participants were ever educated about the formal definition of a computer. We asked the participants to provide us with similarities among these two computers and interestingly they all mentioned the availability of Internet and social media as the common factor between a desktop computer and a smart-phone. Their insistence shows that for digital natives, computers are primarily communication devices that we all use to access the internet, share content and interact with other individuals.

2.6 The Second Generation of Digital Natives

[332] presents a scenario made up of individuals in their early 20s, in the first decade of the 21st Century, who make use of mobile internet and refers to them as Digital Natives. These individuals are surely digital natives but one has to be realistic when making such assumptions on the availability of technology. These individuals are digital natives with respect to their exposure to technology but we proposed in Chapter 1 that there is now a second generation of digital natives. These children were born towards the end of the first decade, they have an unprecedented access to technology [285]. A a mobile computer (such as a laptop, netbook, tablet, etc) can be found in most homes. Wireless broadband is accessible around the house thus giving these children the facility to stream high quality content, on any device, from anywhere in the home. Finally, the fact that touchscreen devices are dominating the market, children are finding it extremely easy to use these devices because they only require their little finger. This leads us to the scenario where one year olds manage to master the intuitive touch interfaces of their tablets whilst sitting comfortably in their baby bouncers. This does not mean that they've suddenly become a technological expert. Far from it, but at such a young age, they're already conversing with a computer interface. Apart from this, the amount of information available online which can be consumed freely is also unprecedented. Even the cost of software shot down with the rise of the apps where [246] claims that 90% of the apps downloaded were free in 2013 and this trend is expected to keep on growing. The parents of

these 2DNs are normally open to technology because they were the first generation of Digital Natives. They understand technology and its potential, so they don't want their children to miss out.

In Section 2.3.1, we emphasised that as per current definitions of Digital Natives, one concludes that the way these groups think is directly related to the technology they have at hand. This line of thought sheds light on the way the 2DNs thinks.

In previous sections one could see that Digital Natives think in a networked fashion and reason in random access just like when they use hypertext. On debating the differences between Digital Natives and Digital Immigrants one can also argue that Digital Immigrants cannot simply reason in this way because they were not brought up in the context of hypertext. Same logic applies in the way 2DNs thinks. This new generation was brought up in a context of total mobility, with mobile internet properly available, Wi-Fi widely available, a selection of mobile devices from phones to tablets. It is therefore easier for this generation to think about information in the context of location. It is also easy for this generation to think about how such information is available on different devices and they expect it to adapt itself accordingly. These differences usher the initial differences between the first generation (as defined by Prensky [254]) and the 2DNs proposed in this book.

Unlike the second generation, the first generation was not brought up in a context of mobile data and devices. The first generation is used to wired devices in order to be connected to the information highway. One may also start asking about how will this generation think differently from the previous generation. While we aim not to further tangle the 'Digital Natives Terminology', it is helpful if one also thinks about the first generation of digital natives as immigrants in this new paradigm shift.

By making use of Prensky's metaphors, it is easy to think about this new generation that is being exposed to wireless technology in the following way. The first generation of digital natives was born in a wired technological environment and these are in fact speakers of this technology language. Then, there were the Immigrants who were not native speakers of this technology but joined onto this new land. Now imagine that there is a new technological 'land', the land of mobile technology. This is the home of the 2DNs and this second generation is a native speaker of this mobile technology. Therefore, the other generations or groups landing on this land are immigrants in that context and also need to adopt in their own ways. It is needless to say that the first generation would find it relatively easy to adapt when compared to the original Digital Immigrants.

A generation is always easier to understand once it i superseded more than once. Same applies for user groups. Once we start thinking about the second generation of digital natives, our minds will relax and understand the reality of Digital Immigrants even better.

Our understanding of user groups has to be as fast as our understanding of new technology. In the longer term, society can't afford to understand the effect of technology decades after the realisation of the said technology. This might eventually provide us with very useful trends which would provide us with a better understanding of the situation. It might ultimately boil down to a different attitude towards technology and the same attitude that differentiates a digital native from a digital immigrant. This attitude would therefore lead us towards the ultimate acceptance of technology where through experience and exposure, any user would ultimately be comfortable with using the same technology. Nevertheless, while digital natives might then find it easier to handle the technological environment in which they were born and immersed right from their very first day, digital immigrants might need a stronger and more dedicated approach towards technology in order to foster the same skill, or wisdom.

While it does take decades for empirical studies to take place and make sense, society has to learn from the patterns of these events and prepare itself for the next paradigm shift and learn the skills that would be needed for the next leap.

Chapter 3
Paradigm Shifts

3.1 Introduction

From time to time, the world come across realities that simply cannot be ignored. These realities are sometimes conflicting with the current philosophy when these occur, thus facing reluctant acceptance. The change in the way of thinking is known as a paradigm shift. This term was coined by the science philosopher Thomas Khun. Kuhn states that scientific advancement is not evolutionary, but rather a series of peaceful interludes punctuated by intellectually violent revolutions and in such revolutions one conceptual world view is replaced by another" [176].

In retrospect, one may argue that the information presented in any paradigm shift existed beforehand. In reality, a paradigm shift may be seen as a way of perceiving the same old piece of information but in a different way. Yet, when we perceive this information differently, it enables us to reason differently about the problem at hand thus possibly leading us towards a solution. Agents of change enable communities to move forward from a paradigm to the next.

Paradigm shifts do not only happen in science. Agents of changes that we find around us drive us towards such shifts. A recent example of a quasi-social paradigm shift is the concept of recycling that was sparked by the increasing awareness about environmental importance. Today, we take the concept of recycling for granted and we do think twice before disposing of different waste in the same container. Paradigm shifts also happen in the life journey of every human being, in some cases more dramatically than others. The older one grows, the more experiences one goes through. As we live through experiences, we polish the way we see and perceive situations around us. Below we present a fictitious scenario of two generations of a family going through a major shift in their lifestyle. This will eventually help us to understand how major shifts and changes affect our skills.

3.1.1 A Life Story

Let us for a moment put aside the thought of computers and technology. Imagine we are breathing the fresh air in a barely inhabited area in the middle of a mountainous

© Springer-Verlag Berlin Heidelberg 2015
A. Dingli and D. Seychell, *The New Digital Natives*,
DOI: 10.1007/978-3-662-46590-5_3

area. In this area, there is a cottage where a couple lives. The cottage is isolated and the next building is a 5 minute drive away. The closest grocery shop is 10 minutes drive away while the closest pharmacy is around 20 minutes away. To get to a superstore, they have to drive 35 minutes.

This couple is used to forecast its needs and plan in advance the stock that they need for the next days or week. The lack of proximity with other buildings, commercial services and people forces them to develop a particular approach that would help them survive and live in a relatively comfortable manner. They communicate with their relatives using different technology and every now and then, they visit some friends who live around 40 minutes away. Twice a month, they also travel to visit their parents.

They have to plan and optimise every single step since any miscalculation would result in a considerable inconvenience.

One day, the couple decide to move to a metropolitan city, around 200Km away from their cottage. This time, they opt for an apartment at the heart of this main city. Everything is so close. All they need is to take the elevator and they are just a short walking distance away from all their needs. On the same street, they find a superstore, pharmacy and also a cinema. This time, it is not just them and some other dozen of individuals sharing the same squared kilometre. The streets are busy with cars and pedestrians and one cannot fail to notice the number of pedestrians going around. If you feel lost or need to find the closest shop, you just ask someone walking past you. The need of forecasting, planning and optimising errands is now diminishing.

This couple now has a child, brought up in this environment. Everything is so close. At an older age, he starts noticing his parents buying their daily needs from shops closer to them and occasionally, they go together to a superstore to purchase items they need for a longer term. To some extent, his parents still practise the idea of planning and optimisation, simply because they had been doing that for all their lives. Nevertheless, the growing boy does not perceive this need of planning and forecasting.

Why should he optimise his chores when his friends live so close, the cinema is in the same street and game store is round the corner?

Compared to his parents, he is more agile and adapts quickly to changing scenarios. He is used to an environment that changes quickly. The cinema one day closed down and the owners converted it to a shopping arcade. He quickly learnt from other young people in the sports ground about another cinema close by. He then just had to take a 3 minutes ride on the metro and get to the next cinema 2 stops away from his home. He is used to being well connected and have all sort of information and services at his finger tips.

This story aimed to highlight the paradigm shift that was experienced by a generation and its implications on the next. This paradigm shift brought about new skills in the new generation while reducing the importance of other skills. It is therefore very important for any employer to understand the paradigm shifts his employees experienced in order to be able to appreciate better their skills and utilise them efficiently. It is therefore useless to expect particular skills from younger generations, simply because those skills were there in the previous generations.

The key factors are knowledge of skills resulting form a paradigm shift, making sure to trust the individuals with these skills and carefully drive a sound policy that coordinates and monitors the previous 2 factors.

3.2 Notorious Paradigm Shifts

Society experienced numerous paradigm shifts and in some cases, the rate of change was drastic and relatively fast. Physics is a field of research that experienced a significant number of paradigm shifts over the past thousand years. Below follows a light case study of the key paradigm shifts in Physics Mechanics. The aim of this case study is to help us train ourselves in zooming out and appreciate while identifying the major changes in a research field.

3.2.1 Paradigm Shifts in Astronomy - A Case Study

Due to its grandeur and complexity, astronomy has from the early days of human existence fascinated societies. Prehistoric temples and later the pyramids have links to astronomy in their planning and architectures. This strongly influenced the Greek and Roman societies in their culture and religion. Figure 3.1 below concisely illustrates the major milestones that defined the developments in astronomy that also led to philosophical paradigm shifts.

With their limitations, the ancient Greeks had a celestial model that was based on altitude. This vertically layered model, placed earth and humans on the first bottom layer. Subsequent layers followed with the sky, sun and stars. However, it was Claudius Plotemy who presented a refined model and introduced the concept of a solar system. The importance of the Ptolemaic system was that it could explain the motion of celestial bodies, thus helping in understanding the structure of the solar system [229].

The Ptolemaic model was uncontested for nearly 1500 years until the 'Copernican Revolution' took place in the early 16th century. This was a significant paradigm shift in Physics. Previously, the Ptolemaic model of the heavens explained that Earth was the centre of the Universe. Nicholas Copernicus then presented the heliocentric model where, conversely, the sun is placed at the centre of a solar system with planets orbiting circularly around it. This started a revolution in the thought about matter. The greatness of this paradigm shift is proved due to its effects, that went beyond effecting the thoughts about astronomy [174]. His model shocked the way Western civilisation thought and therefore faced long delays and strenuous resistance [31].

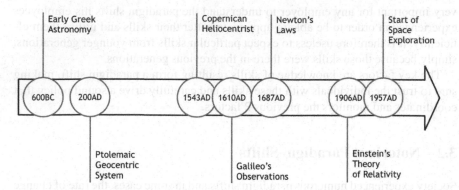

Fig. 3.1 Various milestones in astronomy that lead to notorious paradigm shifts

A subsequent change to the Copernican Model was presented by Johannes Kepler. Kepler's model was based on the previous one and refined the motion of planets around the sun shifting from circular to elliptical.

The Copernican Revolution/Paradigm Shift, was later supported by other important scientists. With the use of a telescope for astronomical observations, Galileo Galilei supported Copernicus' claims by observing the phases of Venus and the moons of Jupiter [90]. In a challenging time for those opposing religious beliefs, Galileo refuted ideas that the sun and planets orbited the Earth, becoming himself a heliocentrist. With such evidence, Galileo attracted powerful oppositions that eventually added more weight to his claims.

In the late 17th century, Isaac Newton sealed the Copernican Revolution with his book 'Philosophiae Naturalis Principia Mathematica'. In his book, Newton provided sound explanation and mathematical models of how gravitation keeps planets in their very orbits [173]. His contributions are still relevant today and uncontested in their context.

The explanations put forward by Newton provided a sense of stability in Physics and the way it developed [175]. This sense of stability also prevented the respectable scientist Lord Kelvin from considering future paradigm shifts. In 1900 he famously said "There is nothing new to be discovered in Physics now. All that remains is more and more precise measurement." This was simply and unknowingly five years before Albert Einstein published his paper on Special Relativity, thus bringing forward another notorious paradigm shift. In his paper, Einstein explained how motion is relative and time should no longer be considered as uniform or absolute. This overturned the concepts of uniform motion. This paradigm shift therefore established that space and time can no longer be considered in isolation in the understanding of Physics. It subsequently meant that time depends on velocity and contraction became a fundamental consequence at appropriate speeds [328].

The Second World War pushed technological development in various fields such as aeronautics and rocket technology. The Cold Ware pushed the limits of space travel and as a result, humanity experienced the first significant milestones in the late 1950s. The realisation of space exploration brought around the latest paradigm shift, which triggered the curiosity of many scientists and led them to explore what was till then, theoretical. Other individuals in different fields started to dream about science fiction, aliens and other related topics. All of this encouragement to keep on enquiring and exploring resulted in physical and mental liberation in further scientific and philosophical exploration.

3.3 Paradigm Shifts in Computing

Right from their early days, computers played a very important role in society. They were at first astonishing machines and later it was curiosity that drove many to ponder how these can be effectively applied, anywhere. This section goes through the major developments in the history of computing affected the way users interact with machines. It also explores the events that allowed for wider distribution of the concept. With every shift or landmark event, users were required to develop new skills and in some cases, previous skills were no longer required.

3.3.1 Going Electrical

The first machine that made use of a set of instructions to perform a task, was the looming machine created by Joseph Jacquard in 1801. This machine was configured by punch cards. Each card had a sequence of punched holes representing a pattern, the machine would sense the respective pattern and it controlled the threads in the looming machine. A few decades later, Charles Babbage partially created the first mechanical computer that was able to calculate the difference of two numbers [133]. He eventually planned "The Analytic Engine" that would compute numbers according to algorithms inputed through punch cards.

The birth of the electrical computer took place amidst the bombing of the Second World War. The first electrical computer, Colossus, was built by the British mathematician Alan Turing. This computer was used to crack the German Codes and decrypt the messages that were encrypted by the Enigma Machine. Turing's computer served well for military purposes but was never commercialised. On the other side of the Atlantic, J Eckert and J Mauchly were inventing a computer that could be potentially commercialised. The Electronic Numerical Integrator and Calculator (ENIAC) was designed with the thought of having business applications [17]. For the first time in history, a computer was now being marketed to do tasks that humans could do. Their first client was the American Census Bureau which used the Universal Automatic Computer (UNIVAC) which was used to predict the presidential election results of 1952. This deployment was indeed the pinnacle of the paradigm shift since it triggered thoughts on how computers could increase competitiveness and efficiency of entities, being private or public. All these discoveries amplified the relevance of computers. Computers were now capable of carrying out tasks that few humans could only carry out, with difficulty. From then on, people started to look at computers as a machine that could adapt to different circumstances and that can be placed in various scenarios where manual processing was at that time very slow.

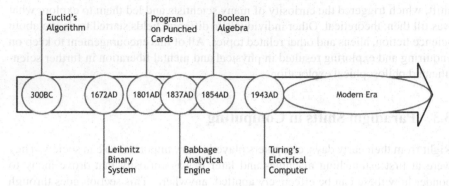

Fig. 3.2 Historical milestones in computing before the electrical computer

3.3.1.1 Information

The flexibility of the electrical computer strengthened the realisation of information processing. This paradigm shift triggered the imagination of many who understood the magnitude of the volume of information that one can possible process. Furthermore, when provided with plenty of information, one is subconsciously engaged in an effort to find the relevant information from that given source. Right after the Second World War, Vannevar Bush was the first one to propose the idea of having an automated method of finding specific information from a particular source [43].

Bush starts his article by saying that in the war, everyone plays his part, even scientists. He continues to explain that while the job of biologists will be unchanged, other fields are developing and creating "strong destructive gadgets" [43]. Bush was here introducing a notion of expanding scientific fields.

In his article, Bush introduces the idea of having a machine helping with indexing information and enabling swift retrieval of this information. He conceptualised this idea in a system which he called *"memex"*, which was conceptually a personal device which stores records, books, all sorts of communication and which can be used to get more speed as an individual. The memex, or personal record, would have been " an enlarged intimate supplement to his memory" [43]. With this statement, Bush ushered the first concept of an information retrieval system.

Few years later, the world started experiencing developments in this new subject. In his recent survey, Singhal gives a brief account of the developments which happened subsequently in this field [290]. Peter H. Luhn developed a system which finds chemicals using punch cards. Later, the same scientists proposed a system which used words as indexing sets and where an overlap with a criterion would result in a retrieval [290]. This was the start of what we today refer to as Information Systems. This notion of receiving information, storing it and subsequently processing it for consumption is today the backbone of all online applications that vary from social to productivity systems.

3.3.1.2 Batch-Processing

The 1950s ushered Batch Processing into the electrical computing scene. This was the first paradigm shift of the computer as we know it today. Batch Processing is an approach that processes jobs one after the other and not starting a job before the current one being processed is completed [298]. This style of processing was used to address immediate business problems such as billing in accounting. Billing is a routine process that has to be finished timely and in a reliable manner. Batch processing was adopted to process such routines and was left computing the data in question until the very end of the designated set. This also meant that the task was dispatched to machine and it was left 'working' until the task was executed entirely.

By this time, society was used to the production idea that was used successfully in the manufacturing industry. Computers were an expensive resource and therefore had to be used efficiently. Batch processing allowed for such processes to be executed irrespective of office hours and just like manufacturing, calculations could also take place overnight. This also allowed for operations to take place without user intervention once the adequate planning and programming procedures were set.

The batch processing paradigm was only effective if users had the proper skills to exploit it. It therefore required users to master their planning skills in scheduling jobs sequentially and properly analyse which jobs should be executed earlier with respects to the application in question. Furthermore, since the processes were not always instantly monitored, one needed to be trained how to properly interpret errors and logging documents resulting from the process. At this point, the required skills-set allowed users to work independently from the computer itself, thus rendering the computer similar to any other industrial machine. This meant that users were impersonal towards the computer since the task of using a computer was simply dispatching a process and feeding it a specific data set that required processing. At this point, users also viewed computer as a threat to their jobs. Therefore, if one did not wish to lose his/her job, the only option was to learn how to use this machine and acquire the required planning skills that were needed to use these new machines. In the meantime, users were also developing significant planning skills that are required in scheduling operations.

3.3.1.3 Time-Sharing

Computers that were available during this and the previous paradigm, were bulky and occupied significant space. The batch processing paradigm showed that a single user was not efficient enough while a group of users actually were. This brought to light a paradigm shift just around one or two decades after the emergence of batch processing.

Batch processing helped numerous users understand the true power of computers. The time-sharing paradigm shift was brought around since it was being noticed that users were sending bursts of information to the computer and waiting for it to process the designated information. It was observed that in the batch-processing environment, it was not efficient to wait for other users' tasks to finish before submitting a new task and therefore, users could be idle while the computer was processing the data set in question.

The Time-Sharing paradigm shift was motivated by this scenario; in other words, having a large machine that only one user could use for a significantly long period of time while occupying significant space. This paradigm therefore resulted in various users connected to a machine using a dumb-terminal, i.e. a terminal with no processing power and acting as an interface with the larger computer. Time-sharing allowed for users to load programs onto the machine's main memory and using it, apparently simultaneously. This is when users started to experience the relevance of more powerful computers and it was the first time that computers were used as a relatively instant tool that could take an input and return an output in a significantly short time. This event created a better link between users and the machines that they were using. Such link could only emerge when users could understand, through experience, the possibilities and limitations of computers and therefore becoming more creative in their utilisation. On the other hand, the skill of planning and scheduling for a relatively longer term that was required for batch processing was slowly becoming redundant.

While there were no particular new skills that the users required to use from the previous batch processing paradigm, security started to be an issue. The need for user authentication and login using passwords was required. Furthermore, system designers had to properly understand the commercial setting and arrangement of a company in order to properly employ systems.

3.3.1.4   Microprocessor

In the early days of electrical computers, around the year 1945, programs were not stored on a computer [240]. On the other hand, today, the concept of memory is part and parcel of the architecture we are used to. Innovation and development of computer architecture recorded significant improvements a couple of decades afterwards. During the early years of the 1970s, we experienced the development of integrated circuits (ICs). The original integrated circuits were modules of interconnected electronic components where only few transistors were connected, generally by soldering, in their first generations [298]. This introduced the modular nature of electronics and it was providing significant improvements (around 25% to 30% every year) to mainframes and microcomputers [240]. It is only imaginable, how this significant change in processing power, continued accelerating with the usage of computer systems in different domains.

This improvement was suddenly accelerated and brought to a new dimension in the late 1970s with the emergence of the microprocessor through the use of semiconductors. A microprocessor is an advanced integrated circuit with tens of thousands of transistors compressed in a very small scale [298]. The microprocessor did not only save significant space when compared to ICs but also provided a significant 35% improvement in performance [240]. Microprocessors could also be mass-produced and this brought down their price and at the same time, accelerated their development.

In 1965, Intel co-founder Gordon Moore, observed and predicted that the number of transistors in an integrated circuit will double every 2 years [225]. Despite being suggested prior to semiconductors, the law is still valid and was proved right throughout the years. Above all, Moore's [225] prediction generated further interest in computing and made people look forward to future improvements of the hardware that subsequently resulted in more realistic and useful software.

Users were not directly affected by the emergence of the Microprocessor. Computation and accuracy were not affected. However, microprocessors did change the way computers looked like because of the considerable processing power performed by a silicon chip of a relatively very small size. This meant that vast spaces were not longer a requirement for powerful computers. It eventually meant that powerful portable devices were a realistic possibility and the positive effect is currently being experienced by all forms of portable devices ranging from laptops to smart phones and wearable computers.

This paradigm shift started to challenge the way people were thinking about 'where' to find computer systems. While computers were originally bulky and required significant space to operate, with the introduction of the microprocessor, computers were significantly smaller in size and achieved higher performance. With smaller machines achieving much more than the previous colossal ones, one could also start considering employing computers in further circumstances where space was limited and computers had to be lighter in weight if they were to be considered for applications such as flight.

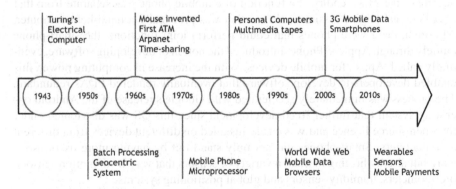

Fig. 3.3 Modern milestones in computing after the electrical computer

3.3.2 Networking

After decades of experience using computers, people refined the way they reasoned about computers. The computing paradigms that were previously explored shed light on the next step. The Time-sharing paradigm shift brought around the notion of having terminals physically separated from the main machine, save for the cable connecting them. This meant that there could be different computers working together while not physically being too close to the same machine. This is in fact the basic philosophy of wired networking and its first examples.

Connecting computers together was a new area of study that changed the lives of millions around the world. In the time-sharing environment, internal wiring was simply used to connect a terminal to a mainframe. Hardware requirements however started to change when it was needed to connect different computers found in different buildings namely through the Public Switched Telephone Networks (PSTN). Whilst this connectivity was relatively slow, it was the start that society needed to move towards the networks we know today.

With the widespread of networked computers, configuration started to change. The security concerns brought around through Time-Sharing still applied in the context of a wider network. Furthermore, new knowledge about the hardware needed for the creation of a network was required in order to effectively disseminate this useful technology. However, this ultimately resulted in a very positive widespread, particularly after people became aware of the potential brought forth with the connectivity between machines over significant distances.

3.3.2.1 Mobile Devices

The paradigm shift brought forth by Mobile Devices was the driving force behind the Wireless Computing paradigm shift. Mobile devices were the ultimate enablers of this concept and took forward the idea of having wireless computers. Following the significant advancements seen in the first decade of the 21st Century, smartphones have indeed become more than just a phone. Until the late years of the first decade of the 21st Century, the function of a mobile phone was separate from that of a Personal Digital Assistant (PDA) that was practically a hand-held computer. Although, there were phones that could perform joint functions, the smart-phone namely through Apple's iPhone, introduced the notion of developing software, eventually called 'Apps', for mobile devices. With the increase in computing power, this enabled developers to develop software that eventually rendered PDAs redundant. The process was immediately accelerated with Google's release of the Android operating system that further freed development since this OS was distributed under the open source licence and was easily installed on different devices from different manufacturers. Smart-phones did not only stand out by having more usable software but also by hosting in them a variety of sensors that vary from motion sensors, thermometers, humidity sensors and global positioning systems.

Smartphones were eventually more affordable and therefore started to replace the previous generation of phones. From then on, people started to reason that the mobile phone is more than just a wireless replacement to fixed line phones with an extra capability of sending short text messages. The introduction of a usable graphical user interface that did not require the use of stylus eased the integration of the device with our day to day routines. The notion of simple, straight to the point applications also revolutionised the way we reason about software. After a few years with smart phones and their significant developments, users started to expect even more functionality on these phones. It is now expected that any software that runs on a traditional computer such as a desktop or laptop should also run on smart phones and tablets. This means that one of the major paradigm shifts brought around by mobile devices is the wide-spread availability of software on devices of different sizes thus expanding the idea of 'where' to find and use computers.

3.3.3 The Online World

The context of the initial networks that were briefly explored in Section 3.3.2 was limited and most of the time meant connecting computers in a confined space. In itself, this marked the start of the Internet. The Internet did need these initial networks since it is essentially a collection of networks that together span across the whole globe [241]. It is also important to clear a basic misconception at this stage. The Internet is the physical network or infrastructure that connects different networks, and thus computers together. One can easily compare it to the plumbing. On the other hand, the World Wide Web is different from the Internet since it is about the information flowing in the plumbing of the Internet.

3.3.3.1 The Web

The first web followed what Berners-Lee and Groff proposed in 1992 [23] and this is explored in detail in Section 4.3. The idea of having a group of scientists sharing their results around the world, suddenly evolved into a global reality. The concept was that of a group of global Internet users contributing to a global source but in a very limited manner (when compared to the frequency of the content being queried). In this web, users were mostly consumers of information.

The concept of the Semantic Web was also developed by Berners-Lee *et al* in a vision of having a Web of machines capable of understanding the web data which they would be handling [25]. As its name implies, the Semantic Web deals with the meaning behind the knowledge, with the way in which knowledge is represented and how it is shared once it is efficiently represented [27]. The semantic web differs from the traditional web since it deals with the understanding and the navigation of the actual information by the machines rather than the presentation of information for human consumption [152]. While the consumers of Web 2.0 and the previous web were humans, agents will be the primary consumers of the Semantic Web. In this scenario, machines will be able to "comprehend and process heterogeneous information in a human-like manner" and would thus understand content in context [152].

The term 'Web 2.0' was coined by Dale Dougherty and Tim O'Reilly from O'Reilly Media 10 together with Craig Cline from MediaLive at a conference in the year 2004 [226]. In their paper *'Teaching Web Development in the Web2.0 Era'*, Wand and Zahadat claimed that in around 2003, people started to use the web differently and shifted from the "read-only" environment [321]. Wamelen and de Kool describe the Web 2.0 as a "metaphor for new Internet technologies and applications" [316] while concurring with the shift described by Wand and Zahadat. They also go one step forward and describe Web 2.0 as a "revival" and a "second generation" of the web.

In practice, all the web users experienced the shift described above. We were all used to the web as simple consumers by reading, filling in forms, messaging and using the Internet from a fixed location. This shift introduced the concept of participating as well while using the web as end user and thus becoming co-producers [316]. Furthermore, this also meant that day to day tasks could be carried out online, including certain payments. This also raised security issues with regards to the amount of personal data people were giving out on the internet. The Web 2.0 also gradually enabled us to publish content with more ease and putting the user at the centre of the web [226] by making content dependent on the users thus being more dynamic[130].

3.3.3.2   Cloud Computing

McCarthy had a visionary consideration of computers in the 1960s when he once said that they will one day become a public utility that can be accessed like any other utility. Digging deeper into this statement, one can easily understand that McCarthy meant that computing power would be housed somewhere else and accessed from any other computer, or device, connected to the designated grid. This concept, eventually coined as Cloud Computing, became a reality with the distribution of the internet, thus enabling this notion of computational space and power.

The cloud metaphor aims to abstract the internet into a single centralised platform or a central 'pool of computers' that allows 'peripherals' to store content on it and use applications that are not running on the said peripheral [337]. The properties of cloud computing together with its various possibilities is discussed in detail in Section 4.3.2.

The powerful and wide notion of cloud computing provides users and engineers with different views of the concept [157]. When users realise the reality behind this concept, particularly the fact that one's data is replicated on another machine somewhere over the internet, security awareness issues start to arise. The cloud stores and processes data, therefore users are now becoming more aware about which data is actually being handled away from their computer. Cloud computing also introduces the notion of running application online, via a browser [157]. Users are now skilled at finding the web application that allows them to do what a desktop application used to allow them to do, yet, without any locational restriction. At a deeper level, cloud computing also introduced the paradigm shift of accessing the infrastructure and architecture via a remote machine. This means that the infrastructure could be 'rented' according to the needs of the deployment and on a higher level, the same applies for platforms or operating systems. Cloud computing therefore abstracted the tradition notion of having a physically accessible computer on which software was deployed or even used. This means that users now have fewer limitations when it comes to actually deploying any system. It can also be significantly scalable when one considers the accessible hardware and technology that might not have been as available or affordable as it is today. Therefore, the online paradigm shift does not only mean social connectivity and sharing of content but also the utilisation of physical machines distributed anywhere around the globe or wherever there is a connection. Thus, Internet services such could be rendered a better cloud service. They services themselves such as utility computing and software as a service will be greatly increased, thus strongly affecting the way one uses the internet on a large scale or otherwise.

3.4 The Wireless Paradigm Shift and 2DNs

The thought of a wired connection gives us a sense of security and stability. Albeit the comfort and practicality of mobile devices, the thought of running out of battery life is always borne in the mind of the user. On the other hand, when using a desktop computer, one feels more secure and power provision is the least of problems for the user. The fact of having an 'uninterrupted power supply' (UPS) also buries such concerns deeper in the user's mind.

Did we ever ask ourselves whether we just feel secure in this way just because it was the only method we constantly used till now?

For those used with a 'cord' based technology,it is difficult to completely trust a disconnected device. This might also be an unconscious restriction, deep in our mind, since we all depended on a chord for the initial 9 months of our lives. Our very first cry came when we were disconnected from this chord. The very same unconscious fear unknowingly haunts us once again in our technological ventures. The initial fear before going wireless. This might have been a very challenging technological transition for most persons in the world.

Imagine having a fixed line telephone on an office desk, where everyone around you visits you every so often and leaves some paper on your desk. Now consider piles have built on your desk burying the telephone. Following the cord would quickly help you trace the phone and use it. The quest of finding the phone would not be so straight forward if the device were a mobile phone. The absence of a chord would create a sense of uncertainty while looking for the phone. Now, consider how you would feel if you left the mobile phone somewhere else?

This simplified example aims to present the reality that although we find mobile devices useful, we still feel the importance of chords.

If chords are so important for us, why do we still move away from them? The answer is simple. Just like our first step towards freedom was marked with the cutting of the chord which kept us alive for 9 months, going mobile and wireless is our first step towards our technological freedom and therefore birth.

This is exactly how the new digital natives feel.

Availability of mobile devices started its realistic increase in the late 1990s. Everyone born since the early 2000s, is to some extent familiar with the notion of having a mobile device. Just like the first generation digital natives are familiar with the concept of computers, the second generation of digital natives are now familiar with the concept of mobile devices.

The notion of an evolving web and boosting participation through web 2.0, strengthens the users' global communication. The 2DNs are being brought up in a reality where they contribute to knowledge and share content instantly on the web. They are further empowered by doing this anywhere through the use of mobile technology. Cloud computing is also changing the way new user deal with technology.

Trends in this area show there are different views by which users can use cloud computing. These trends [157] [337] show that more investment and development will take place in cloud servers, thus rendering better cloud service. They depict views whereby cloud usage such as utility computing and software-as-a-service will be greatly improved, thus strongly affecting the way one uses the internet on a large variety of devices.

3.5   The Road Ahead

This chapter aimed to introduce the notion of paradigm shifts and their respective relevance to applied computing. The world experienced numerous paradigm that were brought along with different developments in science and technology. These significant developments did not only develop the scientific knowledge and its resultant application. In most cases, these developments shocked the way society reasoned about fundamental principles and therefore progressed accordingly, paving the way for new paradigm shifts to take place. In order to introduce this concept, this chapter started with the traditional case study of paradigm shifts in physics that changed the way we think today. Subsequently, a number of milestones that were achieved with the introduction and development of computing, were explored. All these milestones in computing also affected the way the world developed and with it, the way our lives drastically changed in less than a century of the electrical computer. In the coming years, we are expecting other major paradigm shifts. In fact, [115] claims that in 20 years time, 50% of the occupations we know today will no longer exist due to the rise of Artificial Intelligence and [137] prophesize that AI might even take over the human race. However this comes to no surprise since we've been there before with the technological revolution in the 80s when people thought that computers would make everyone redundant. We've been there with the industrial revolution. Even Aristotle in 320 BC as can be quoted in [91] thought about it when he said "If every tool, when ordered, or even of its own accord, could do the work that befits it... then there would be no need either of apprentices for the master workers or of slaves for the lords". Notwithstanding this, [250] reminds us that this situation is almost inevitable now, but its up to us to combine our resources with our technologies and give them a vision which benefits everyone.

Paradigm shifts in computing were not only bound to the actual machine but also to the way different machines could be connected together with little limitations. This interconnection coupled with the ever decreasing size of machines in question also lead to how computers seamlessly blended our daily lives with ubiquitous computing becoming an emerging reality that cannot be avoided. It is the way that devices are nowadays communicated, without any physical chords that is even leading the new generation of digital natives to think differently (to what we traditionally did) about the use of computers. With new interconnected devices of significant computing power, the questions digital natives have to answer are 'Where to find computers?' and then once found, 'How to use these computers?'. These questions which explored in further details the various dimensions of our reality, are blended together using the technology that we have available today.

Just like the way achievements throughout the evolution of mankind marked different eras, computers are also marking today's day and age. Stone age, Bronze age and so on are related to pre-history but the age of semiconductors, fibre and wireless connectivity are defining the world as we know it today and with it, groups of people utilising this technology in their very own way.

Chapter 4
Blended Realities

4.1 Introduction

Emerging technologies and their applications are moulding the way in which digital natives are developing. Technology is affecting people's lives. However, we also know that it is strongly affecting the way organisations function and develop. Digital natives will soon be the employees joining companies and soon afterwards, driving them. We also understand that we need to blend technology and operations with people. A good blend can result in a flourishing business. This is however not just limited to businesses. Our daily operations are also affected in the same way and it is therefore very important that there is a holistic understanding of blending technologies.

This chapter broadens the scope of what was traditionally perceived as blended realities. The continuum of Milgram et al [220] is used to explain the different ways of how technology is categorised between total virtuality and total reality. However, there is a link between the actual devices that enable modality of realities with the actual applications that are used within any reality. In this chapter we will explore the main ingredients that can help us better understand how realities are actually blended. The idea is that once these three concepts are blended together, software ideas can be generated thus fitting the needs of digital natives. In Chapter 6 of this book, we explore how the process of building software can be adjusted to better understand the new users, the new digital natives.

This chapter first explores these three realities in isolation:

Devices
> This section will briefly explore the notion of different hardware that is making it possible to create new software components capable of handling different situations.

The Web
> The world wide web ensures the continuous flow of information around the globe and with millions of devices (of different types) being connected to the same web, information is taking new forms. This is furthermore expanded with the notion

© Springer-Verlag Berlin Heidelberg 2015
A. Dingli and D. Seychell, *The New Digital Natives*,
DOI: 10.1007/978-3-662-46590-5_4

of having devices, uploading localised information which they collect from the various sensors.

Applications

Once there are devices that provide different experiences (while collecting information that is then delivered via the web); it is up to applications to give proper context to the information collected and create new experiences. Applications can take different shapes and in this chapter, we will explore the relevance of gamification and serious games together with productivity applications.

This chapter is then concluded with an exploration of how these realities can be blended together. This section will show how reality and virtuality can be bridged and properly employed to optimise the processes we have today. The most significant enabler of blended realities is indeed ubiquitous computing. This will be explored with respect to these three realities. There are also new trends arising while this book is being written that prove the relevance of blending all of the above together.

4.2 Devices

The reality of digital natives cannot be discussed without first discussing the factor that brought to life this reality i.e. devices. When one discusses digital natives, intrinsically we have to discuss the devices which these individuals use. Devices are also the showcase of technological progress and in a way, a measure which represents this progress. This section briefly explores the technological evolution which gave life to this generation of digital natives. Devices will be categorised in a way which will reflect the different generations of digital natives.

The discussion and exploration of the new digital natives is directly related to the evolution of the devices that happened during their days. The way that the new digital natives reason about technology and exploit its use is naturally inspired by the devices they had at hand while growing up. One of the most remarkable attributes of technologies in the beginning of the twenty-first Century is the wireless aspect in technology. In this book, we relate the evolution of digital natives with this technological characteristic that strongly influenced the way we interact with technology. This characteristic also brought to light new ways of using traditional devices and in other cases, lead the way to the evolution of previously known technological concepts. Coupled with the developments in hardware development, the first decade of the 21st Century also saw the integration of devices and concepts.

These developments happened just during the childhood of the new digital natives that we are exploring in this book. For these individuals, wireless and compact technology is an absolute reality that is taken for granted by many of this generation.

4.2.1 Gaming Platforms

This section explores the different hardware used in the gaming environment that is significantly popular with the different generations of digital natives. The term 'platforms generically refers to the hardware on which the game is played [11]. These systems vary from Desktop Computers, Mobile Devices to more dedicated hardware such as Gaming Consoles or Portable Consoles and each contributes to a particular gaming experience. Each platform carries with it a particular measure of immersion in the virtual environment and provides a deeper experience if it is for entertainment purposes or other serious situations such as education and training.

At this point, the link between software and hardware is at its peak importance. Game developers follow a set of coding priorities in order to ensure that optimal performance is achieved in the final product. The considerations that are taken into account from a software perspective are speed, size, flexibility, portability and maintainability [270]. This section explores different gaming platforms. Each platform is discussed in terms of its relevance to gaming in general and also in terms of its contribution towards the user experience in relation to these five priorities.

Most of these properties tend to vary according to the class of gaming platforms for which games are developed. However, there are a couple of properties that are more generic. Flexibility and Maintainability are factors that are directly related to the quality of the code used in development and are therefore dependent on the software process and not on the target platform as such [270].

Computers and consoles are gaming platforms that are relatively restricted in physical portability. This category includes desktop, laptop computers and also Gaming Consoles such as Sonys PlayStation, Nintendos Wii and Microsofts Xbox. These platforms are combined together for the purpose of this chapter since they share most of the properties considered in such a class of devices. Speed on such devices is generally taken for granted. Nevertheless, one has to be careful with such a conclusion since although they offer plenty of processing power during rendering, the quest for further speed on these platforms is a difficult challenge. The processors and operating systems of these machines tends to be complex and non-deterministic at a low level [270], therefore making it close to impossible to optimise the instruction level. The size of the game is not a priority on this platform due to the resources which such devices have. On the other hand, the majority of the space taken by the game is used by the artwork and the audio while the actual programming takes just a small percentage of the game space [270]. Code is also generally portable between different platforms that belong to the same classes [270].

Portable gaming platforms, mainly Mobile Devices such as smart-phones, tablets and Portable Gaming Consoles such as Sonys PSP and Nintendos GBA are increasing in their popularity, especially in the ever-growing reality of casual gaming. Generic restrictions for development on such portable devices are the screen size and the limited battery power [288]. Design for games used on portable consoles is restricted to the low-powered processors of such machines [270]. Games executed on such devices are carefully coded to dedicate processing power to the actual game and reduce speed in minor modules such as menus and help.

Mobile devices offer a variety of new opportunities in the gaming industry. Processing power in mobile devices is ever increasing and this is accompanied with an array of sensors that can read important values [84] that vary from location to temperature and humidity. Further more, due to the gaming platform hosting such games, it is possible to integrate these casual mobile games into the daily routine of players [131]. Games are generally easily transferrable between the leading mobile platforms, i.e. Apples iOS and Googles Android. Game engines such as the Unity engine allow straightforward compilation routines for each of these platforms, thus making the code as portable as possible [71]. On the other hand, online games pose the challenge of compression when it comes to transmitting larger volumes of information [270]. Such dependency on the network (in order to download content during gameplay) should be kept to an absolute minimum, since the interruptions during transmission should not result in disruptions within the game.

4.3   The Web

The public use of the Internet is always increasing, turning it into a continuously updating data source. This section aims to provide background about the Web and the way it is used. After analysing the evolution of the web, this section provides information about technologies and techniques used today on the web.

The World Wide Web (WWW) was proposed by Sir Tim Berners-Lee and Jean-Francois Groff in 1992 in their paper named *"WWW"* [24]. Berners-Lee and Groff designed the WWW concept for storing records and linking them accordingly. In their paper, they explain that the WWW runs on every machine once a browser is installed and explicitly describe it as "a generic information retrieval engine" [24]. The WWW eventually developed as a collection of web pages accessible through the Internet.

The first web followed what Berners-Lee and Groff proposed in 1992. The concept was essentially that of a group of global Internet users contributing to a global source in a very limited manner (when compared to the frequency of the content being queried). On this web, users were mostly consumers of information.

4.3.1   Web 2.0

One of the key success factors of Web 2.0 was the "leverage of customer self-service and algorithmic data management to reach out to the entire web" [233] which is also sometimes referred to as the *"long tail"* allegory. This contributed to the success of Web 2.0 since the service of applications in this paradigm get better as more people make use of them by participating [233].

Another breakthrough of Web 2.0 is that it made people make more use of the Internet. Hailpern *et al* claim that the dynamic evolution of Web 2.0 applications "have blurred the line between desktop applications and the Web" [149]. This reality also resulted in a struggle for software markets since it made the Web a functional platform by sharing numerous attributes with the other platform, the operating system [233]. Nonetheless, the web has various advantages when compared to an operating system. These are namely; the absence of ownership, open standards and cooperation agreements between companies, which make the latter cheaper for the end user and developer [233].

Many people are also making use of Web 2.0 applications in an effort to improve the general wellbeing of their society. In fact, many governments are making use of Web 2.0 by setting e-Government as a priority since they understand the far reaching capabilities of this web [307]. In their paper *"Web 2.0: A Basis for the Second Society?"* Wamelen and de Kool explore the Dutch scenario of e-Government while acknowledging that such an approach makes information more accessible [307]. This e-government solution using Web 2.0 was also shown by Papathanasiou *et al* in the Cypriot environment [213]. Similarly, various efforts are in place to utilise the benefits of Web 2.0 in education and learning [146]. We will explore this topic in further details in Chapter 8.

As case in point, Colomo-Palacios *et al* developed a web based platform employing Web 2.0 and the Semantic Web to create a social network dedicated to e-learning in software engineering [146].

4.3.1.1 Products of Web 2.0

In this section the main services of Web 2.0 will be explored and particular attention will be given to the survey conducted by Yakovlev [335]. Many initiatives have exploited the properties of Web 2.0 illustrated above and have triggered significant use from the general public. Below follows a list of these initiatives which characterise Web 2.0.

Weblogs which are more popularly known as **Blogs** are websites which allow threaded discussion of a particular topic [335]. Incidental topics attract the attention of various group members [340] and thus allow for wider ownership and ease the distribution of content [335]. Weblogs are sometimes considered to be in the grey area of Web 2.0 since most viewers are consumers of information but on the other hand, they allow the general public to comment or participate on a particular topic thus touching the properties of the Web 2.0.

Wikis are websites which enable collective collaboration, on the development of content, by members of a community while having the same community vetting the entries [335]. A popular Wiki is Wikipedia which is a website which allows entries about every topic. Yakovlev [335] reports that some organisations are using wikis to collect input from various sections of the same organisation and thus having this as a tool acting as a placeholder for content. There are other wikis which are dedicated to a particular topic.

RSS stands for *"Really Simple Syndication"* and is a medium through which frequent updates of a particular web source are published. RSS feeds are based on XML technologies and may be read by most web browsers and RSS readers [335].

Social Media is a major development of Web 2.0 which has evolved into a field of study in itself. This has expanded after the wide scale integration of Web 2.0 and its enabling of users to contribute directly to web content [236]. It is a new dimension which has brought plenty of motivation in the creation of new applications but at the same time, opened the doors for various issues. The social web entails online platforms or website which allow individual who have an account or profile to post content [340]. This can be in the form of various products which were developed as a result of web 2.0. The use of these products promotes the sharing of information related to personal activities and the expression of ones' ideas.

Multimedia also found its place in Web 2.0. Technologies such as Podcasts and Vodcasts are a way of broadcasting audio and video to a set of subscribers via an aggregator [335]. Web 2.0 also allowed for users to upload their multimedia content on websites such as YouTube[1] and easily share it online.

4.3.2   Cloud Computing

Cloud Computing is a model or infrastructure of running applications on interconnected different devices over the Internet with computing resource sharing as a primary aim. The term computing resources ultimately includes storage, processing power, applications, servers and also other networks. As briefly mentioned in Chapter 3, the cloud metaphor aims to abstract the Internet into a single centralised platform or a central configured 'pool of computers' that allows 'peripherals' to store content on it and use applications that are not running on the said peripheral [337]. These online resources can then be released and provisioned with minimal management efforts through interfaces to facilitate the work of service providers [245].

This expandable infrastructure is a product of grid computing [338] and a model that goes beyond the actual sharing of resources which was traditionally known as client-server or mainframe infrastructure. The American National Institute of Standards and Technology (NIST) [245] provided an organised way to understand cloud computing by identifying a set of characteristics, service models and deployment models. These attributes and aspects of cloud computing were also acknowledged or covered in other works [338] [44] that are contributing to a harmonised understanding of the subject area.

The main characteristic of cloud computing is its availability through proper connectivity by making use of interfaces that eliminate the need of human interaction to access any of the resources offered by the cloud. Access to the cloud is not limited to a particular modality and is therefore accessible from computing devices such as desktop computers to mobile devices such as tablets or smart-phones. With such a range of devices being capable to connect to the cloud, the cloud model is ever-growing and is also developing an elastic infrastructure [245]. The on-demand

---

[1] YouTube was launched in 2005 by Chen, Hurley and Karim: http://youtube.com

notion of cloud computing also brings into the scene utility computing [44] that entails the characteristic of a measured service [245]. In other words, cloud computing is used as a utility in the same way we access physical utilities such as water and electricity. The 'consumption' of services from the cloud utility is measured and pay-per-use models may also apply. This is then further abstracted since the usage of resources on the cloud are provided by different entities who are pooling in the same cloud.

Once a cloud infrastructures is characterised by the above, different service models can be employed. The NIST identifies the three main services that are briefly explained below. These are stacked in the following way: The first layer presented is the provision of Infrastructure on which any customised platform or software can be executed. On top of this, one then finds the provision of ready-made customised platforms that serve different purposes for the deployment of custom applications. Ultimately, on the top layer, one finds the software that can be used and accessed directly to process any data required by its users.

Infrastructure as a Service (IaaS)
> This aspect of the cloud involves the provision of services that are normally associated to hardware. This virtualised access to computing power, storage space and network allows users to deploy applications that require particular computing resources [245] that might not be immediately available to the same users. Furthermore, this also contributes to more portability of the applications in questions particularly because it allows for customised operating systems to be deployed on an infrastructure of choice.

Platform as a Service (PaaS)
> This service provides a platform or operating system available on the cloud. This varies from operating systems to programming language execution environments to web servers. It strongly facilitates the deployment of applications at a very low cost since the owner of the application has to focus on the quality of the application without the monetary or time investment in setting up the infrastructure and platform in question. The end-users using the application would be ultimately using third-party platform [338] without a particular effect on their experience.

Software as a Services (SaaS)
> This layer is at the highest level of abstraction and provides the service of ready made applications that users or entities can use with the same characteristics of the cloud. This would include the idea of pay-per-use software that would be too costly to develop. A practical example would be the use of ERP or HR management systems available on the cloud for the use of small enterprises [338]. It would enable the users to use the applications from the web or a program interface without any consideration about the maintenance of the software, management of the utilisation of computing resources or even the operating system upon which it is running [245].

These services are becoming more available and this is due to the different types of deployment models being adopted worldwide. These models vary in terms of physical deployment and location of the cloud [44]. It is therefore understandable that deployment models vary from private clouds to community and public clouds. There also exists the possibility of a combination of these thus resulting in a hybrid deployment model [245]. Private-clouds are only accessible to a single organisation but it might be managed and/or hosted by a third party entity off-site. On a slightly broader scale, one can also find community clouds that are similar to private ones but with access limited to a group of organisations. The most accessible clouds are public-clouds that consist of an infrastructure owned by an organisation and then made available to individuals or other organisations. Hybrid infrastructures consist of a combination of the other three deployment models bound together by technology that caters for portability and instances of load balancing between models [245].

4.4   Applications

Software is what ultimately renders computer useful and usable. Furthermore, as explained in Chapter 6, software needs to also be tailored to the tasks users need to perform. In this section, we will explore the major categories of software. In recent years, games have been developed not just for entrainment purpose but also for serious applications. In fact, they are being used as an innovative way of engaging users for a relatively longer time while managing to achieve serious tasks. On the other hand, the us of other more serious software such as productivity applications is also very important in today's work places. This section therefore explores how these two major categories are developing and how they will affect the way the new digital natives will ultimately use computers.

4.4.1   Games

Gaming is a daily reality for many individuals which is being experienced over different channels and media [203]. Games do not only help individuals to relax from the daily tedious routine. They also act as a social opportunity for individuals to engage within "social groupings [153]. There have been various concerns about the effects of games on individuals [343] but at the same time, there are counter arguments that demonstrate that games play in important role in our modern society [214]. However [102] presented a longitudinal study whereby the consumption of video game violence is correlated with youth violence rates of the past two decades and the results show that there has been a decline in the youth violence rates, contrary to what was being envisaged.

4.4.1.1 Game Genres

Just like any other set of products, video games are carefully segmented. Nevertheless, games are segmented by their style of interaction and gameplay rather than just by their narrative or visual differences [11]. Since the definition of a genre is bound to the interaction style, it is also appreciated that genres are evolving in a pace relative to that of technological advancement. Genres are also important for the day-to-day classification of games on a social level. Each of these genres may be adopted in different scenarios to train or entertain digital natives and even other audiences. Gamified software and procedures taking the shape of any of these genres would probably be more effective than normal software.

The gameplay of action games requires good coordination of different parts of the human body together with the respective mental processing effort. Cinematic components in these type of games are integrated within the gameplay and take advantage of the sub-genres mentioned below [11]. Other types of games such as adventure games are sometimes placed in this genre since they require rigorous coordination and at the same time, are not strictly bound to a backstory [227]. For example, adventure games share plenty of properties with action games but focus much more on the story [270] and quests throughout the storyline.

The gameplay in First Person Shooters (FPS) offers a viewpoint of the players vision [11] within the virtual environment. FPS games are closely associated with the military due to the nature of their gameplay, game content and game mechanics [34] These games are known to influence the player due to their restricted narrative structure that is particularly experienced during single-player mode. In such a playing mode, the user interacts with fictitious characters within the game and accompanies them in combat according to a well defined storyline [34].

Third-person games inherit all the properties of an action game but similarly to FPS, the difference lays in the interaction style. This sub-genre is distinguished by the setup of the 3D camera within the game that allows the player to see the character or avatar on the screen [11]. Assassins Creed is a game that can belong to this sub-genre.

Simulation games are those that allow the player to go through an experience within a realistic environment, that is not directly bound to a narrative [227] and which, respects a certain set of rules. This set of rules includes physics, sports and also music. As their name implies, racing games are those in which the player participates in a racing competition. This varies from car racing to air and sea vessels. Games in this genre include those that either follow a real world competition such as Formula 1 or fictitious competitions that are based on fantasy such as Super Mario Kart. Such games are often known for their physical realism and are sometimes accompanied by a backstory [227]. While racing games can be also regarded as sport games, it is generally understood that sport games are those that simulate a traditional sport such as soccer or basketball. This genre carries forward the competitive aspect of tradition that is in itself a pull factor. Through multiplayer infrastructure, sport video games are becoming more popular since they exploit this competitive psychological aspect of gaming by having individuals competing on a global scale.

Games can be used to simulate a certain rhythm or musical piece and challenge the user to interact with the designated work of art. This sub-genre brought around innovation in interaction and gamers can interact with such games using motion-detection software and also dance pads. Such games have positively impacted the music industry by boosting sales in certain music genres when linked directly with games belonging to this genre [124].

Turn-based strategy (TBS) are characterised by the restriction of achieving a task, such as for example controlling the majority of the game area, by having each player play a single turn at a time in a round robin fashion. On the other hand, Real-time strategy (RTS) games do not progress incrementally in turns and the actions within the game happen instantly and continuously [48]. RTS games can pose a resource production challenge to the user. In this case, the player has to gather resources in order to proceed with production of units that would allow progression within the game. Another challenge which RTS games pose is known as tactical battles where the player has to use military units to gain control of area on the screen and defeat the enemy force [48].

4.4.1.2   Benefits and Concerns of Games

Games can be used to positively influence individuals about their lifestyles and in certain cases it was also proved that games steered certain individuals away from drinking habits [203]. Game are also within themselves an opportunity for individuals to engage within a group in society and get in touch with other individuals [153] particularly in situations where individuals are sometimes detached from society.

Studies [248] [214] show that by the end of their teenage years, individuals spend tens of thousands of hours online and this compares to the same amount of time an individual spends in class by that age. This figure includes the time spent using computers and games. Such a figure is also a concern for most adults [343]. However, Jane Mcgonigal [214] states that this is potentially very profitable for society and because of this, young people should be encouraged to spend more time playing games.

4.4.2   Productivity Applications

The use of mobile devices by digital natives is not limited or restricted only to games. Efforts are constantly being made to transform different processes that were traditionally carried out using other devices into usable software for mobile devices. This varies from word processing applications to customer relationship management applications. When excluding games from the choice of categories in the application marketplace such as Google Play, Microsoft's Marketplace or Apple's Appstore, one is spoilt for choice when it comes to choosing applications. Although productivity might refer to a particular category of applications, for the scope of this book, we are using this term to refer to applications that help the user carry out a real life task such as a word-processor to write a letter or a social media application to connect with other users.

As explained in Chapter 2, digital natives are finding various ways of using mobile devices. Efforts by companies producing platforms for these mobile devices are also being made to facilitate development of applications by digital natives themselves and in fact it is not the first time that senior figures of such companies boast about the number of young people who are coding applications on their platforms.

In Chapter 6 we will explore how digital natives can be involved in the production of applications however this is not solely limited to their involvement in the software development life cycle. It is important that whether digital natives are involved in the development of applications or actually building them themselves, they find proper and effective use of the said applications in order to improve their efficiency in the everyday tasks that they carry out.

Earlier populations of digital natives or even digital immigrants were trained on the use of applications on desktop or laptop computers. A straightforward example would be the use of word-processors or spreadsheets packages. Earlier populations were used to the use of these packages using the physical mouse and keyboard with the user interface that was actually designed for such a setup. However, when it comes to using the same package on a mobile device with a user interface designed for a smart-phone or a tablet, a steep learning curve kicks in. The mouse pointer is no longer available to simply move around and popping up comments to facilitate use. Every touch will start committing an action and therefore one becomes pretty much restricted in that. The use of an on-screen keyboard also hinders the experience with the actual screen being reorganised every time the keyboard is prompted. Due to the limited size of the portable display, the user interface has to be reorganised in terms of buttons and their respective functionality which makes it pretty difficult for users to get used to it.

On the other hand, once a person gets used to productivity software on mobile devices, the possibilities are vast and can improve the work of most people from students to journalists and other professionals. Nevertheless, the new generation of digital natives is not spoilt with the use of physical hardware when running such applications. They use mobile devices intuitively and with proper care, these applications can be developed in a way to be better understood by this emerging generation and thus boost their productivity.

An interesting experiment was carried out during the research phase of this book. A protocol analysis exercise was conducted with elderly people (around 70 years old) who had never used a computer in their life but were given a tablet as a Christmas present. Most of them recalled that at first, they started using it reluctantly not to disappoint the person who gave them the gift. They also installed and set up a Facebook account on their tablet and they started using it. They explained that after a couple of demonstrations, most of them were capable of using it on their own without the need of further help. During the exercise, some of them also recalled how in the past they had briefly tried to use a desktop computer but could not get hold of the basic interaction skills such as the use of the mouse. However, it was also explained that since the tablet only responds to where you actually hit with your finger, it made their life easier. It also resulted from a number of them, that the simplified user interface of Facebook that focuses on graphics accompanied by simple text such as 'like' or 'comment' made it easier for them to grasp.

This experiment was also very useful when it came to better understanding why the new generation of digital natives is responding to the way they are utilising mobile devices. Just like these elderly, they were never exposed to a mouse or a keyboard so their first point of contact was actually the mobile device. Learning how to use the mouse would then be just a skill like riding a bicycle or eventually driving a car for convenience sake.

Efforts are also being made to carry out productivity tasks related to education, tourism or even virtual reality sets. The wide use of devices in different contexts is motivating end users to better engage with computers and make human-computer interaction a better reality since they are motivated to use the said devices to carry out tasks that previously could not be easily carried out.

4.5   Blending These Realities

This section deals with the different ways we interact with applications. It starts by exploring the mixed-reality continuum which ultimately differentiates between the degree of real and virtual content in an experience. The relevance of augmented reality together with virtual reality will be also investigated in this subsection. Subsequently, the notion of ubiquitous computing will be presented in a separate subsection. Ubiquitous computing is the availability of interconnected devices all over the place and the way they behave together.

When presenting these two powerful concepts, this section demonstrates how the blending of all this available technology together with the ever-improving applications will lead towards a new revolution in computing which is currently being experienced by today's users. This means that this interconnectivity of concepts will be the norm for digital natives in the near future.

The benefits of this blending varies but above all, a resultant benefit is the increased efficacy of software through better immersion, which ultimately results in an improved presence with regards to the experience offered. Immersion is related to the technology providing the experience and not to the actual user experience. It provides the degree to which the system is involved and can be measured using the following five levels: Extensive, Surrounding, Inclusive, Vivid, and Match [293]. On the other hand, presence is the feeling obtained from the surrounding space when it is sensed by the user in order to rate its closeness to reality. This is not about how accurately the space mimics real life, but rather how close it feels to real life [29]. It is the feeling of being physically present in a specific space, albeit actually being situated in another [333]. The primary factor in presence is focus or rather complete focus on the virtual environment stimuli. This is a very similar experience to what we experience when reading a book, watching a movie or even day-dreaming.

4.5.1   *Virtuality-Reality Continuum*

The reality-virtuality continuum presented by Milgram et al is a spectrum of mixed realities having the real environment at one end and the virtual environment at the other end as illustrated in Figure 4.1. The aim of their original paper was to classify the relationships between augmented reality and the larger class of technologies which are referred to as Mixed Reality [220].

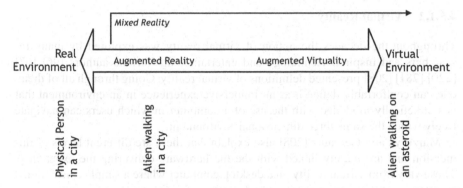

Fig. 4.1 Reality-Virtuality Continuum (Adapted from [220])

The leftmost end of the continuum represents a completely real-world environment with no virtual artefacts whatsoever. In this part, the world in question is completely un-modelled [220] which in practice means the physical objects around us in real life. On the other extreme, at end of the continuum one finds a totally virtual environment where all elements and objects are artificial. In this case, the world is completely modelled [220] and the designer may choose to either replicate the objects that appear in the real world through computer generated graphics, slightly modify them or even create objects or scenes that are in no way related to the real world.

In the middle of the continuum, one finds worlds that partially model experiences through Augmented Reality of Augmented Virtuality. Augmented Reality is the most relevant from the central part of this continuum particularly because of its increasing availability through smart-phones and mobile devices. Virtual reality is also demonstrating an increase in its popularity and relevance as discussed below.

4.5.1.1   Augmented Reality

In principle, augmented reality is the process of having a computer capturing a real-world scene by making use of, for example, a camera and superimposing computer generated graphics onto the captured real image or stream. Augmented Reality can be experienced through different types of displays that range from head-mounted displays, hand-held devices with displays to also spatial displays that are located at a relative distance between the user and the real scene. The application and popularity of Augmented reality through hand-held displays was boosted through the availability of powerful mobile devices. In our previous work [83] [81] we explored different applications of augmented reality that vary from travelling to edutainment. However, the new digital natives would be experiencing head-mounted devices, such as Google Glass, as an ever-growing norm in the use of Augmented reality. With further improvements in the field of computer vision and data compression, the relevance of augmented reality will be further appreciated due to the ease of blending content over transmission.

4.5.1.2   Virtual Reality

Throughout the decades the notion of virtual reality was explored by many researchers and inspired technology and entertainment. Various authors [96] [67] [220] [281] [208] presented definitions of virtual reality. Going through all of them, one can comfortably define it as an immersive experience in an environment that is synthetically modelled with the use of a computer in which users can navigate freely within the same three dimensional environment.

Marzuryk and Gervautz [208] also explain that there are different levels of immersion that are mainly linked with the the hardware delivering the experience. These vary from virtual reality on a desktop computer where a simple conventional monitor is used up to an improved adaptation of a fish-tank environment where stereoscopic viewing is allowed. The most immersive experience of virtual reality is the one which uses head-mounted displays that can support sensory output and input which is closely linked to the user.

Applications of virtual reality vary from visualisation of data for productive tasks to entertainment such as games. Cultural and education applications such as virtually visiting sites that are not easily reachable or which are no longer available are also on the rise. Teleoperation of remote devices is also a potential field which will also provide significant results particularly in dangerous situations such as bomb-disposal and the handling of hazardous materials. Besides these seemingly traditional applications of virtual reality, a new emerging application will probably boost this field of study. The acquisition of Oculus VR by the social media giant Facebook is indicating the blending between the virtual reality onto what we today understand to be social media. This could eventually evolve in a fully-fledged three-dimensional social media environment similar to Second Life[2] with real life content uploaded by the users of the network. Such a development would bring into play a major development on the current user interface of social media applications that closely resemble an electronic noticeboard.

4.5.2   Ubiquitous Computing

What's the time?

Effortlessly, you probably just looked at your watch, your tablet or phone time. You might have also checked the time on your computer or perhaps the clock on

---

[2] http://secondlife.com

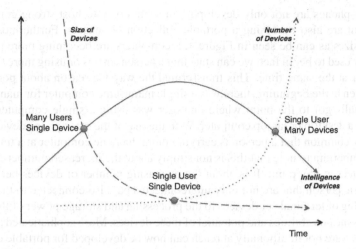

Fig. 4.2 Relationship between the size of devices, their number and intelligence against time (adapted from [206])

a wall. These 'devices' where there before you were reading this. However, you probably did not think much about them in this context. Devices of different forms and shapes, mechanical or digital, that provide us with an indication of time are actually all around us. Yet, we do not really notice them until we need them.

This renders clocks as ubiquitous devices since they are practically everywhere, anywhere. There might also be the misconception of equating ubiquitous with invisible devices. However, these devices are in reality blending with their environment, thus, generally unnoticed.

Computers have come a long way until we are now finally discussing realistic ubiquitous computing. Up to a few decades ago, computer performance was not taken for granted. The probability of running a task on a computer without actually crashing was not very high. Once computers stabilised and became more available in every home, questions started to rise about how to connect these computers together. This was realised after Berners-Lee [24] contribution with the creation of the WWW. It was from then onwards that information was adequately shared over the internet. All this allowed information to flow.

Subsequently, society experienced the mobile reality where phones that were simply designed to call and SMS slowly started to become computers themselves. The emergence of the smart-phone brought significant computer power to everyone's pocket. Individually, smart-phones were powerful and eventually, mobile networks started to complement such performance. While technology was to some extent available, it was not always affordable. It is now being noted and acknowledged that smartphones are becoming more available and more software is being developed for such devices.

Smart-phones are not only developing in such a way to host strong computing power but are also becoming a portable collection of sensors. Furthermore, with smaller sizes as can be seen in Figure 4.2, computers are becoming more portable then they used to be. In fact, we can state that a person carries or using more than one computer at the same time. This transformed the way we reason about portability since when in the beginning, there was a single mainframe computer for many users, it eventually got to the point where one user was using a single computer which could be a desktop or laptop computer. With the rise of the smartphone devices, it is now very common that a person is carrying more than one computer at a time.

It is important to note that this is not simply about the increased number of available devices at any point. It is about the increasing number of devices being used by a single person that are not only interconnected but also connected to the cloud. The linking of terminals or devices to the processing and storage power of the cloud increases the possibilities and potential of these devices. More sophisticated applications that were not traditionally at reach can now be developed for portable devices. In this effort, intelligence and storage is being shifted towards the cloud. This reality is pushing towards an emerging generation of relatively dumb terminals. These mobile devices are now being required to fetch and upload data and content to the cloud with little processing being carried out locally.

Other examples of 'dumb' devices vary based upon their context. A simple example would be the computer inside a car that detects a telephone call from a mobile device and facilitates the call through the use of hands free functionality by using the microphone and speakers of the car itself. The job of this onboard computer would be to connect (via wireless technology) to the smartphone which would be lending its cellular antenna and to a few other components for the call to be delivered together with the onboard computer. The onboard computer that uses a less-visible microphone in the car together with the speakers that are usually used for entertainment purposes which might be hidden or considered invisible. It is in fact blending nicely inside the car, cleverly using existent components of the car.

Another example would the use of an intelligent earpiece computer which would connect to the smartphone during calls or for instance with an entertainment system inside an office. Even though this might be a small device, it is still not invisible. The device in this second scenario is definitely smaller than the one mentioned in the first scenario. This is exactly what is leading many people to argue that devices are becoming widespread and invisible. This line of thought is today's interpretation and product of Moore's Law. A range of blending devices might be perceived as invisible since they would properly fit their context. Buxton ** explains that this is similar to walking in and out of a bathroom without using anything and when out again someone asks you whether or not you noticed the toilet. Most people would probably say no. However, if one was to walk in and out of a dining room and find a toilet there, being asked the same question on the way out would surely yield a different answer. This is because the object in question was out of context in the dining room.

4.6   Future Blending

Devices, the Web and Applications. We normally perceive these as separate elements but rarely dedicate appropriate time to ponder on how they are linked and what actually links them. As shown in this chapter, these components are easily and vastly linked through different processes we carry out without knowing. Furthermore, emerging technology and approaches are providing us with new ways to utilise these three realties in a more blended approach.

Gone are the days when these were not seen as one. These are being blended with the continuous development of the notion of Ubiquitous Computing that is making devices available anywhere by being connected anytime together. Furthermore, from a user's perspective these realities are being blended with approaches such as Augmented Reality and Virtual Reality that were traditionally seen as distant but that are becoming speedily available on affordable and high-performing devices at everyone's reach.

The ease of blending these realities also comes to play with the openness of programming applications for the leading mobile platforms. Companies owning these platforms together with others developing middleware software to facilitate development are striving to encourage users to code their own applications. It is also interesting that older groups of digital natives are being very active in this development. The development of applications on mobile devices together with the interconnection of various enabling technology is meshing these efforts (that were traditionally separate) into one holistic endeavour.

Chapter 5
Nurturing Digital Natives

[47] claims that education is there to guarantee that all students, irrespective of their background can benefit from the learning experience to the full. This is achieved by ensuring an active participation in the community. To do so, when considering our fast changing world, educators must move away from old methodologies and experiment with new ones in order to allow their students to reach their goals. This does not mean that the introduction of technology in the classroom is a guarantee of success. Far from it. As Tanya Byron puts it:

> The technology itself is not transformative. It's the school, the pedagogy, that is transformative.

There have been various experiments conducted using different pedagogies in conjunction with technology but the most effective one seem to rotate around play and fun. Play is something which we do ever since we are in our mother's womb. [317] associates the pleasure associated to rocking on a playground swing to the gentle swaying motion of the womb. Others mention sound and light games from within the womb as being the basis of adult play. Because of this, educators need to look into the use of games in order to help them reach their goals in an effective way. Let's not forget that play is something enjoyable and engaging so if they manage to infuse these qualities into our educational system, play will revolutionise the way we learn.

5.1 Learning Styles

Before delving further into play, let's first understand how people learn[239]. Extensive research has been conducted on the concept of learning styles and several models have been proposed and applied. The idea behind it is that a person will learn better if information is presented in a manner which he is able to process. [73] states that the view of educators and academics who have studied the differences that exist between individuals learning styles is that everyone has a personal preference for receiving and processing the information that is presented to them. In the past, this

© Springer-Verlag Berlin Heidelberg 2015
A. Dingli and D. Seychell, *The New Digital Natives*,
DOI: 10.1007/978-3-662-46590-5_5

was difficult to achieve since the teacher could not handle all the students in the class individually. With the introduction of technology in the classroom, this can be achieved by allowing each and every student to make use of different learning styles (which can be provided by their personal device). The student might prefer to listen to the lesson using his headphones, maybe watch videos or even go through some interactive exercises. The choice is ultimately of the child. According to [63], there are around 70 learning styles and they have been grouped into families as can be seen in Figure 5.1.

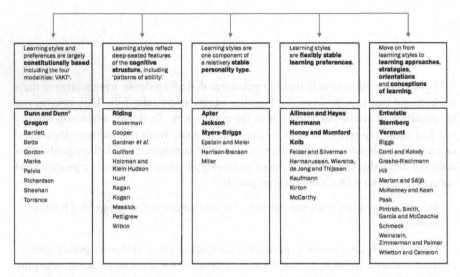

Learning styles and preferences are largely **constitutionally based** including the four modalities: VAKT².	Learning styles reflect deep-seated features of the **cognitive structure**, including 'patterns of ability'.	Learning styles are one component of a relatively **stable personality type**.	Learning styles are **flexibly stable learning preferences**.	Move on from learning styles to **learning approaches, strategies, orientations** and **conceptions of learning**.
Dunn and Dunn³	**Riding**	**Apter**	**Allinson and Hayes**	**Entwistle**
Gregorc	Broverman	**Jackson**	**Herrmann**	**Sternberg**
Bartlett	Cooper	**Myers-Briggs**	**Honey and Mumford**	**Vermunt**
Betts	Gardner *et al.*	Epstein and Meier	**Kolb**	Biggs
Gordon	Guilford	Harrison-Branson	Felder and Silverman	Conti and Kolody
Marks	Holzman and	Miller	Hermanussen, Wierstra,	Grasha-Riechmann
Paivio	Klein Hudson		de Jong and Thijssen	Hill
Richardson	Hunt		Kaufmann	Marton and Säljö
Sheehan	Kagan		Kirton	McKenney and Keen
Torrance	Kogan		McCarthy	Pask
	Messick			Pintrich, Smith,
	Pettigrew			Garcia and McCeachie
	Witkin			Schmeck
				Weinstein,
				Zimmerman and Palmer
				Whetton and Cameron

Fig. 5.1 Learning styles families according to [63]

Kolb

David Kolb is the person recognised as having started the modern learning styles movement when he published [171]. His original and later work ([170] which has six distinct features) was based on the work of several other twentieth century scholars namely John Dewey, Kurt Lewin and Jean Piaget amongst others. He states that:

> "Learning is the process whereby knowledge is created through the transformation of experience. Knowledge results from the combination of grasping experience and transforming it."

Someone who is an effective learner needs to learn using the following abilities: concrete experiences, reflective observations, abstract conceptualisations and active experimentation. These models cause conflicts within the learning and it is by resolving these conflicts that the learning occurs. Eventually, the learner will start preferring one of these methods over the others. There are also four different learning styles according to Kolb which are the converging style, the diverging style, the assimilating style and the accommodating style. As part of his techniques, his team also developed an Adaptive Style Inventory used to measure flexibility in learning and a Learning Skills Profile used to assess the skill development in the four skill areas mentioned earlier [199]. Throughout the years, several other theorists (such as Honey and Mumford in [141]) used Kolb's original ideas to create their own questionnaires.

VARK

VARK[1] is a model created by Neil Fleming which was published in [104]. The idea behind it is to create a dialogue between learners and teachers when it comes to learning. The questions used are based on real life experiences aimed at challenging the learner in order to reflect on the ways in which they deal with information, how they process it and how they present it. When the person answers the questions, a result is obtained for each VARK category thus revealing the person's dominant preference. The result does not necessary single out a single preference but a person can have multiple preferences as well. A critical aspect of VARK is the fact that it only provides feedback on a person's communicating means and to really understand the learning style of that person, more analysis will be required. On the other hand, this model is useful to learners who would like to tune their delivery towards the needs of their classroom.

Honey and Mumford
The work of Honey and Mumford [141] was based on Kolb's original model but with an alternative learning style questionnaire. Their focus was also shifted towards managers and management activities thus making it more relevant for the administration of an organisation. By using this study, managers could focus on their strengths as learners and thus help them develop a more all-rounded approach to learning. The main difference with other questionnaires was that it asked about the general behavioural tendencies rather than about learning behaviours. Through their studies, four learning styles (Activist, Pragmatist, Theorist and Reflector) were identified which are very similar to Kolb's learning cycle [305] thus verifying previous findings as well. Honey and Mumford also believed that although a person has learning preferences, these preferences may be adapted and improved upon.

Felder and Silverman

The research of Felder and Silverman looked into the learning difficulties experienced by engineering students and professors. They proposed in [100] various learning styles based upon previous work by Kolb, Jung and the Myers-Briggs Type Indicator [211] hoping to resolve these issues. Felder eventually revised

---

[1] Visual (V), Aural (A), Read/Write (R) and Kinesthetic (K).

the proposed dimensions leaving four dimensions in total. [101] eventually designed an instrument aimed at identifying the learning preferences based upon these four dimensions. The system has fundamentally two applications, first and foremost its an academic exercise into the various learning styles present within the classroom which would help educators to devise lessons which target the various learning needs. Secondly, by using the questionnaire, students can identify their strengths and weaknesses so as to further enhance the learning process in their favour.

Although there have been various studies on learning styles, there are other schools of thought. [239] argues that although people give out their preferences in these studies, they might only be preferences and nothing more. He even goes further to state that educators do not need to give too much weight to these preferences because in reality, it is unlikely that an educator uses only one learning style with all his students. Further still, it might be beneficial for a learner to try a new learning style which his not accustomed with. According to [73], this alignment (between learner and the consumption of information) is just an assumption which we make. Thus, a positive approach would be to present the learner with different learning styles and let them decide which suites them best.

5.2   Game Based Learning

From the previous section, it is evident that children need an environment where they can experiment and enjoy doing so using different learning styles. A process where failure is not seen in a negative light but rather as an important stepping stone for learning. The most conducive paradigm which exhibits these properties is the one associated with gaming. Children are already used to play and in the past couple of decades, we have seen a further move towards digital gaming. First of all digital video games are more engaging thus giving the player the thrill associated with the experience whilst allowing him to be in control of the game. This illusion of full freedom is very important for a child because they have the impression that they are breaking out of their real life restrictions (which were imposed by their guardians). Well crafted games give the players that adrenaline rush from the comfort of their armchair. They allow the child to progress at his own pace thus making the learning task much more manageable. This is very different to what actually happens in class whereby the teacher stirs the group based upon the average abilities of the children. Thus children whose abilities are beyond that average tend to get frustrated because they classroom experience is not engaging enough. On the other hand, children who are way below the class average might find it hard to catch up with the rest of the class. A game can provide students with a personalised program thus helping them achieve their goal at their own pace. These achievable steps will give the user a positive sense of accomplishment which is way different than traditional schooling. Accomplishment isn't measured at the end of a school term or during the yearly exams but these systems normally provide continuous assessment. Further still, if one fails these assessments, he is directed through a path which ensures that the child understands the underlying concepts and resubmits the test as soon as he is ready. Students don't have to repeat a whole year any more and the notion of failing becomes part of the learning process whereby a fail is not the end of the world but is always followed by another chance to succeed. Therefore, it comes to no surprise that 97% of teens play video games [186]. A recent study which we conducted not only reinforces this figure but also found that 97% of children aged between 3 and 12 years of age also play video games. This can be directly attributed to the introduction of the touch interfaces which makes it much more easier for small children to use digital devices. The popularity of video games is unprecedented; in 2009 the UK video gaming industry outperformed Hollywood movies for the first time ever and [113] predicts that it will reach the $82 billion mark by 2017. Notwithstanding this, there are still industries which look down at video games as if they are inferior. [51] claims that some film critics throw insult at films by comparing them to video games. Some adults still view games as being a waste of time which could be employed to better use. However this seems to be changing since in our research, we found that 57% of the parents with teenage children play. On the other hand, 71% of parents with children aged between 3 and 12 tend to play at home. Thus, in the eyes of the parents, the value of digital video games is slowly increasing and this will in turn effect the way in which they look at gaming. People are finally understand the benefits of games and this has been proved over and over again. Studies such as [19] and [164] claim that games help the gamer practice inductive reasoning, increase visual skills, improve the integration of information from diverse sources and allows the players to tackle cognitive conflicts. The level of realism being reached by powerful gaming consoles such as the PlayStation or the XBox is unprecedented. These systems are pushing further the barriers associated with graphics, animation, sound and user experience. In this past, century, what started as a sophisticated calculator turned into a powerful simulator. In fact, there are very few things (if any) which cannot be simulated on a computer.

Pilots use simulators to learn how to control their complex vehicles. Doctors use medical simulators to learn about the human body and the procedures which should be applied to different situations. Soldiers make use of military simulators not only to train but also to familiarise themselves with hostile terrain before they actually go there. Thus, it follows that if computers can be used to train real life professionals, they can also be used to prepare children for their real life duties when they grow up. Playing games to practice skills is vital for any creature capable of learning according to [66]. Originally, games were designed to absorb the player mainly for an economical reason (because gaming companies want to sell their games) but the fact that players are absorbed in a game implies that the game has captured the full attention of the gamer thus the gamer tends to learn more details from the game. We can claim that learning is a constant in each and every game, with or without the player's awareness. This is not bound by time, in fact when a child starts playing, he also starts learning. The fact that video games are part and parcel of a child's experience (from a tender age) implies that new skills are being developed which were quite rare before. These include the handling of different channels of information simultaneously as per [255], in fact, this can be considered as being an important trait of the 2DNs. Gaming improves important hand-eye coordination and makes the child sharper. This has also been confirmed in [272] which claims that games enhance the psychomotor skills of the gamers. [105] goes further by stating that everything we do, even making a cup of tea, rewires our brain, let alone the prolonged exposure to the internet and games. We have also seen earlier that different people learn in different ways, some are auditory learners so they need to hear the lesson, others are kinaesthetic learners so they learn by doing and the list goes on. But the truth is that in traditional classroom learning it is very hard to cater for all kinds of learners. On the other hand, gamified systems can do that quite easily thus offering a personalised approach to learning. Examples in [181] and [212] clearly show that serious games increases the motivation of students and their desire to learn. This is not just attributed to the fun factor experienced when playing games but there's also an element of curiosity where the player is happy with his achievements and intrigued by the forthcoming steps. By setting contained tasks, clear goals and achievements as specified in [183] education can easily motivate students to reach higher goals.

When one analysis the lives of digital natives, [214] claim that on average, young people play around 10,000 hours of games by the time they reach 21 years of age. [118] goes further by stating that 10,000 hours in a person's life is enough to create an expertise in a particular field. Thus, these children are not only expert gamers where gaming becomes a fundamental part of their life but they also excel in skills found in games as can be seen in [237]. It seems clear that the current education system was designed for a different generation of students, definitely not the ones we have today. In the traditional system, students compete for the top grades and there is seldom any group work. Games on the other hand promote teamwork (especially the social ones) whilst still retaining a healthy competition, a feature which is very desirable, later on in the workplace. This cooperation happens through intricate communication which is quite normal in Massive Multiplayer Online Role Playing Games (MMORPGs). In these games, players participate in team based missions so they are forced to look at problems from a group's point of view and they have to understand that each of them has a specific role which is vital for the success of the mission at hand. Different games allow students to explore scenarios which would not have been possible inside the class. The Sims 3 has been used in [42] to study and analyse what it means to be homeless and raise a child. SimCity 2000 is used in the classroom [6] to teach geographical concepts especially those related to urban and spatial planning. Civilization [49] can be used in order to highlight historical milestones through both diplomacy and wars. Students will not only debate the political implications of their decisions but they can also try them out in the game. This gives the students an insight into what it really means to run a government and therefore, if they decide to pursue that career, they will already have some experience.

Even though the benefits of having game based learning in the classroom are very evident, most teachers seem to be quite reluctant to implement them. First of all, some of them feel that the traditional approaches have been tested for ages and they have been accepted by the community in general. Because of this, they are considered safer, so they prefer to stick to them. Secondly, parents might not appreciate these new approaches and thus pressure teachers to use more traditional methods. Even though these new approaches have been tested in schools such as the one described in [276], their implementation is not straight forward and it requires careful planning from the teacher's point of view. Games must be tailor made for specific syllabi which would obviously require an additional investment from the school. Even though off-the-shelf products already exist (such as commercial games), they do not normally hit all the aspects required by the syllabus and most of the time, they tend to veer off topic.

5.3 The Gamified Classroom

Since mainstream games are not up to the job (because they were not designed with the syllabus in mind), they cost a lot of money to modify and considering that it is hard to strike the balance between an educational game which is also entertaining, educators are opting to gamify[2] their classroom. Gamification does not necessary involve the use of an electronic system but the introduction of gaming mechanisms in the existent educational system. As described in [341], it might include points, levels, leaderboards, badges, challenges, quests and all the different elements which one would expect to find in a game. In so doing, educators will inject an element of fun in the curriculum thus boosting motivation and increases engagement. The scale of gamification used is up to the educator. He might setup a point system whereby students are rewarded for participating in class or for completing their homework correctly. These points might then contribute towards a progression through different levels. Towards the end of the year, the level achieved by the students might reflect their final grade thus in so doing, it would reward students who where consistently active whilst also remove the stress from final year examinations. The fact that students are aware of their own levels can help to boost competitiveness amongst class mates both within classes and amongst classes. Let's not forget that one of the strongest motivator is undoubtedly peer pressure. When one realises that his grades are having a negative effect on the whole class and as a result holding it back, he is normally more prone to work harder. These approaches are rather simple, they do not need complex implementations, yet they can be extremely effective.

In a survey we conducted amongst 128 teachers starting from primary up to tertiary level, it transpired that most teachers use games when teaching small children. However as the children grow up, these teachers do not make use of serious games or gamification techniques in their classroom. This seems to hint to the myth that games are just for young children and not for adults. In fact, more than half of post-secondary teachers never heard about these paradigms. There is also a big disparity

---

[2] Gamification is the use of game related elements such as leaderboards, point system etc to other domains such as education.

between what the teachers actually think. 90% of those teaching primary and secondary school children believe that games are beneficial for their students. On the other hand, only 23% of those teaching elder children have this belief. This seems to imply that when students make the transition from secondary to postsecondary education, the time for games comes to an end as they enter adulthood. However, when one analysis these results under the microscope, it transpires that out of those secondary school teachers who find games beneficial for their students, only 55% actually use them in their teaching and this figure falls drastically at higher levels plummeting to the 4% mark. Luckily, things seem to be changing. The number of adults playing games on a regular basis is increasing and the majority of new teachers are being taught about serious games and gamification techniques. Notwithstanding this, teachers are still faced with the issue of limited resources, thus it will be hard for them to implement some of these techniques without the help and support of from the institution where they work.

We also conducted a similar survey amongst 300 students from primary up to tertiary level. Only 16% of students below tertiary level knew what serious games where and this shows that there is no exposure in the class room for these techniques. On the other hand, 78% of all students believe that games are beneficial in the educational system. This result is not surprising when one considers that a substantial chunk of them are avid gamers so games form part of their day-to-day life. As they grow older, they start to look beyond the entertainment element of games and they start valuing other elements such as the benefits of having serious games in the classroom especially with respect to simulators or real life case studies. Most students (around 80%) seem to agree on the fact that lessons should be enhanced with gaming elements. Students prefer having interactive lectures where they can contribute to the ongoing discussion whilst they learn through hands on analysis of the facts at hand. Notwithstanding this, the level of scepticism amongst students seems to be still high. This emerges from their reply to a question regarding the negative effects of gamification on education. In fact, 40% of the students think that gamifying a lesson will have a negative effect on the educational process.

From this study, it emerges clearly that both students and educators tend to believe that serious games and gamification techniques are beneficial and are definitely a better learning tool than traditional methods. But as we argued in the previous sections, it still is not viable to implement full blown games in the classrooms. To introduce them in the schools, governments must help educational institutes financially so that they can afford them. Also, the industry must develop more of these games and make them more easily available. The most viable alternative seems to be the gamification of the teaching approach. This approach is much cheaper but then one would lose a lot in terms of exciting teaching material (such as simulators, etc) which digital video games can offer. However, the result is still a positive one whereby the teacher can gain the students' undivided attention and make the lessons more interactive than with the traditional approach. After all, these methods too have their disadvantages as well and they are still being tested in various schools. However, with the proliferation of technology in these teaching settings, we expect to see more serious games and gamification initiatives in the classroom.

5.4   The Transition to the Workplace

2018 is the year when digital natives enter the workplace according to [18]. It is imperative for businesses to understand the fundamental differences and the unique capabilities which these new workers have if they want to build a motivated workforce. First of all, these young adults live online. They have been doing so since they were born, they use the computer to play, for their research and now for their work. They have access to the latest technology both in their home and outside. The language and metaphors used in games have been infused in their day-to-day lifestyles especially after playing for approximately 10,000 hours of games in their life. Because of this, the workplace of the future has to be different than the one we are used to today.

Goals

A final aim is vital in every game. It might be a matter of saving the princess who has been stolen by a monster. One might be called to save an entire kingdom. Or even more humble goals like cooking a delicious meal. Irrespective of the game, they all have a goal. But sometimes, goals might look too massive to complete, because of this, games tend to divide a goal into various subgoals which are more manageable. The hero of the game must complete various levels in order to reach the final one. According to [8] if an employer wants his employees to stay highly engaged, he needs to instil into them a sense of progress coupled with meaningful work. This might sound obvious but in most cases, top management seem to be more interested in their long term goals rather than the micro environments within their organisation. This leads to the issue of overlooking the importance of small and daily wins for their employees. They need to structure work into a series of small but manageable units if they want to keep their staff motivated which will eventually lead towards long-term success.

Feedback

When playing games, the player receives instant feedback. This then helps the player adjust his tactics and strategies. If it is positive, then it is used to reinforce his good behaviour, otherwise it allows the player to learn from his mistakes whilst adjusting quickly to new scenarios. In the workplace, what we normally find is a once a year performance review which is too little too late in most cases. Companies need to provide their employees with fast and meaningful feedback thus allowing them to grow and learn at a fast pace.

Transparency

Progress is something which is known in a game. A racing game might have a map highlighting one's progress vis-a-vis the competitors. A fighting game might have indicators of health and time. All the games measure progress in some way

or another both in real-time and also over a longer period of time. In the workplace, an employee is not given access to this information. Most of the time they
have to query their superior and even in that case, the answer might not always
be straightforward. The measure of performance is most of the time a mysterious
formula which is not always easy to understand. This complicates itself if the
performance measure is pegged to colleagues working in the same office or to
others located somewhere else. Whereas one can roughly gage the performance
of fellow colleagues, it is almost impossible to get an estimate of the performance
of those located remotely. Companies need mechanisms capable of capturing this
data, analysing it and share it with their employees in a simple format which is
understandable by everyone.

Badges

In most social games where it is important for a player to share his achievements with others, they need to be represented using some sort of system. That
is why games make use of badges, trophies and all sorts of representations in
order to showcase the skills, achievements or reputation of a player. In reality, these symbols are valueless outside of their context but within a particular
domain, they are metaphors to represent a very specific achievement. In most
organisations, the specific skill and strength of an individual is normally overlooked because what matters is how good he is at executing a very specific task.
Most of the time, people are chosen based upon their academic achievements
and little notice is given to other attributes such as interpersonal skills. These attributes are extremely important especially when assembling great teams whose
task is to reach very specific goals. Keep in mind that one can easily form a very
competent team but elements such as group dynamics can easily shatter everything. Thus, organisations should build mechanisms into the system which allow
their employees to earn rewards for their skills and abilities whilst providing a
self-validating mechanism which can be conducted on a regular basis.

Levelling

Badges are normally awarded for short term goals but when it comes to longer
sustainable achievements, games normally have levels. A level is a sort of senior-
ity indicator related to games. Level 20 in angry birds signifies that the player
has successfully passed the previous 19 levels, that he has dedicated a substantial
amount of time and energy to do so and that he has build a certain amount of
skill. Levels also help to give a macro perspective of the game because if a player
knows that the game has a total of 40 levels and he's in level 20, then it automatically indicates that he has reached half way. When it comes to the workplace,
it is a known fact that a job for life is a thing of the past. Young people tend to
switch jobs easily, experiencing much more jobs than their parents did (in a life-
time). The problem with these knowledge workers is that when a company loses
one of them, its not the service provided which has been lost but the intellectual

capital which is contained within that person. That is why a lot of companies try to externalise the intellectual capital and digitise it inside a computerised system. To try and limit staff turnover as much as possible, employers need to create a career path made up of different milestones aimed at helping the individual learn and grow whilst enhancing his status in the process.

Onboarding

Most games make the player go through a tutorial level before actually dumping him into action. Some of it is subtle and provide in-game notifications thus guiding the user as he proceeds in the game. Others provide tutorial levels which are specifically designed for this purpose. The process of on-boarding has been highly mastered in games thus providing the user with the information he needs without taking too much time from his game. Different games require different tutorials. Some games might be straightforward to grasp whilst others such as SimCity might have complex elements (zoning, economics, etc). Games will not plunge the player into the unknown but they will hold his hand and teach him what he needs to play the game successfully. In a similar way, employees are not expected to understand their duties through a manual. There are organisations who are investing millions of dollars in computer based training or other forms of e-learning such as the Massive Open Online Courses (MOOCs) but in reality, they are not really successful because only a small percentage of people like using them [98]. Companies need to adopt game like mechanisms in order to motivate their employees towards adopting procedures and increase their performance at work. These techniques will help them reach mastery level where they can get further promotions thus helping both the organisation and the employee to progress further.

Competition

Competition is essential in any game. Players can compete either against non-player characters controlled by the computer or against human opponents and with the advent of the internet, online competitions became much easier to organise. When under control, competition is healthy because it pushes the players beyond their comfort zones and urges them to test their limitations. Organisations already use competitions in their structures and sometimes they can be quite fierce as well when they are competing for promotions and raises. The problem with the current systems is that they tend to be informal competitions, with fuzzy rules thus making it hard for the employees to compete. Also, such a structure is unscalable and cannot be automated if the organisation grows. Because of this, companies should setup clear competitions with well defined rules in order to motivate their employees towards improving their results.

Teamwork

People are social animals, they also have a need for relatedness which means that they give great value to their social connections. Because of this, most of them find it easy to collaborate as part of a team in order to compete against other teams. Being part of a team means that they also have the opportunity to connect with like minded individuals and as a result, bond together. The members of a team work together towards achieving a common goal. As can be seen in [200], this cooperation also happens because of peer pressure since team members strive towards achieving the results agreed by the group whilst avoiding at all cost to be labelled as the weakest link.

Being always connected means that younger people spend a lot of their time on social networks such as Facebook & Twitter. Social networking is the top internet activity in the world and it is still evolving with 60% of social media accessed through a mobile device rather than through a desktop computer. On average, people spend more than 2 hours per day on these platforms. Organisations should leverage on this technology and get their project teams to utilise it and exploit its potential. In fact, these technologies will allow the firm to go further by making it easier to get people from different geographical locations to collaborate together. In so doing, it will drive competition whilst fuelling networking and boosting the knowledge-sharing of the company.

5.5 Beyond the Workplace

Living in an interconnected world made up of networked computers, hundreds of TV channels, video games and all sort of other devices means that the new breed of digital natives are aware of what is happening around them even in far away lands. Most of the time, they are more informed about events happening abroad than those happening in their backyard. This gives them new ideas and opens their horizons thus making them explore new possibilities. Rather then seeking passive entertainment, these youngsters are after the experience being offered. In fact, digital natives want to go deeper, they want to understand the meaning behind an exhibit. If we consider a museum as an example, they want to re-live the story of the exhibits as seen from different viewpoints in order to understand fully what is going on.

To do so, there are various technologies which are being utilised. First and foremost, games are being utilised in all sorts of settings (as we've seen in the previous sections). When a person plays a game, he is immersing himself in a role. During the game play, he assimilates himself with the character being played. He shares the excitement offered by the game, the risks and also the delusions associated with losing in a game. It is very common to have players spending hours playing a single game while losing cognition of time. When one visits a site such as a museum, curators would like their visitors to experience the same level of immersion. To do so, one can easily rely on modern technologies.

The most obvious technology is without doubt the mobile phone. This offers three main advantages; first of all, the entry cost of access to the technology is low since most people own one already thus curators can simply make their visitors download an application on their device before they access the museum. The second benefit is that the visitors are already familiar with their device so the learning curve needed to get used to the application is very mild. This is very different to using devices such as audioguides which are extremely limited in function and they can also be complex to get used to. The last benefit offered by these devices is their level of sophistication. Modern smart phones have various sensors including gyroscopes, compasses, WI-FI, Near Field Communication, dual cameras, GPS, etc. These sensors make it possible for the system to track the location of the devices inside the museum and thus send to the device contextual information. This is very important especially if we would like the museums to move towards personalised experiences. The era of set menus whereby visitors are presented with a finite set of exhibits together with a snippet of information is reaching an end. People want a choice, they want to decide what to see and they want to decide upon the level of detail they expect to view. This can only be achieved via personalisation of the different interfaces which can be adopted such as Virtual Reality (VR), Augmented Reality (AR) or also Holographic Displays (HDs).

A VR experience gives the user the illusion that he is inside a virtual environment. There are mainly two types of VR systems, one where the user has to wear a headset and his entire vision is provided through the headset. The image changes based upon the orientation of his head thus giving the user the impression that he is immersed in this virtual world. The problem with this approach is two fold; the headset can be bulky and obviously reduces the realism of the immersion. The second problem is the head to eye coordination. Since the image does not changes based upon the eye movement but on the head movement, it can reduce further the level of realism achieved by these systems. Notwithstanding this, these systems are relatively cheap and large organisations (such as Facebook) are banking on their mass adoption in the coming years. The other approach does not requires a headset but the virtual environment is projected into a large box or a room which the user can explore. This is normally much more expensive than a headset but the level of realism achieved can be pretty high. Together with the projections, engineers add simple effects which stimulate the other senses and increase the level of realism. If the simulation requires a windy environment, fans can be utilised. In the case of rain, water droplets can be sprayed on the users. Scent printers can be used to stimulate the nostrils thus enhancing further the experience. A close example is the Museo Archeologico Virtuale (MAV - Virtual Archeological Museum[3]) of Herculaneum where the people living in the area were wiped out by the eruption of Mount Vesuvius. The museum provides three dimensional reconstructions of the life in those days thus giving the visitors an unprecedented experience. Apart from various projections, the museum also uses vapour curtains and fans to simulate the eruption of the volcano thus making visitors relive the events of 79 A.D through a 4D experience[4].

---

[3] http://www.museomav.it/
[4] An experience which adds to a 3D film, physical effects such as wind, rain, etc.

AR reached a high level of maturity in the past few years. This was achieved thanks to the rise of the smartphones; whereby the superimposition of virtual elements over live streaming video (from the phone's camera) became possible on the device itself. Mobile phones became so powerful that they could easily analyse the video stream, infer associated content and display it almost immediately. In so doing, mobile devices are capable of showing virtual directional signage thus guiding visitors of a museum whilst also customising their path in order to suite their individual needs. Exhibits cab be analysed by the device and relevant content displayed. This includes text, images, videos, 3D walkthroughs and all sorts of hypermedia. But the experience is not only one way; visitors are not limited to consuming the content being provided but they can also create new content or enhance existent one. In fact users can easily create virtual graffiti or tags on the artefact which can be accessed by other users. In this way, a discussion is initiated between the various visitors thus bringing the exhibits to life. In the near future, AR will take a further boost thanks to the introduction of Glass technology whereby the augmented display will become much more natural than holding a mobile phone in front of the eye.

HDs are the next big thing because they will liberate users from the restriction of owning a device. Augmented information can be displayed on top of any exhibit without the need of smartphones or wearable devices such as glasses. This means that even though the initial cost of such a display might be high, the cost of entry for visitors is next to nothing. The potential of such a technology is vast. Visitors can roam around animated displays thus giving flat artefacts an additional dimension. They can also interact with them in such a way which was not possible before since the interaction would be shared amongst a group of people watching the artefact. If we consider a model reconstruction of a neolithic temple, it can be easily animated through HDs. Visitors will be able to see and understand how neolithic people lived. More than that, visitors will be able to use speech recognition or gesture based interaction in order to dialogue with the virtual individuals, by asking them questions and also getting them to do specific task (such as a request to show the user the sacrificial ritual, etc.). The dialogue will not be restricted to only one visitor but different people can barge in and continue dialoguing with the virtual characters.

This leads us to two other technologies which are currently available. The first one is the interactive table whereby different people can interact with a horizontal flat surface. By doing so, they can flip through various documents, watch videos, 3D reconstructions, take quizzes, etc. The limitation of such a display is its 2D surface thus making it difficult (if not impossible) to appreciate the 3D models in their full glory. The other technology is a virtual avatar capable of speaking with the visitors using natural language. The role of the avatar would be that of a personal guide which would inform visits only about exhibits which interests them. Visitors can ask him questions and delve into topics based upon their level of interest. In so doing, the experiences will be personalised and calibrated towards the needs of the user.

Finally, apart from the experience, visitors like to take away souvenirs which would remind them about their experience. However the quality of most souvenirs is questionable. Most of the time, its all about cheap replicas of artefacts which only offer a faint recollection of what was seen. With the advent of 3D printers, visitors can easily take with them exact replicas of the artefacts they've seen. They can also customise them by changing colour or scale it to fit their needs. Further still, they can do so from the comfort of their homes since 3D data can be easily sent by email and printed within the comfort of ones' home.

Chapter 6
Designing for Digital Natives

With more children being exposed to technology, it is only natural to understand that this will also lead towards the creation of a new type of user group. The development of Human Computer Interaction (HCI) has always been effected by the struggle between the physiological and the psychological abilities of the users. This can be further expanded when dealing with the 2DNs. Children have their own inherent psychological and physiological limitations that strongly affect the software that is being developed for them. The entire development process has to be tailored to these realities.

This chapter revisits approaches in software engineering but with a direct focus because it takes into consideration the effect of digital natives on these processes. While the same generic structure of software development would not be changed, there is a significant set of proposed changes that would improve the process. Such an improvement would mean that professionals would better consider the reality of digital natives while ensuring that these digital natives (in other words the end-users) would effectively contribute towards the inception of more suitable software.

Therefore, applications have to be designed with 2DNs in mind. Dix et al [87] defines design as 'achieving goals within a constraint'. This simple definition will be the guide for this chapter. By the end of this chapter, one would be able to understand how to harness the potential of 2DNs while becoming more familiar with the constraints present in such a process.

Employers expect their employees to be as efficient as possible. This does not stop at the point of expectation. While it is important that employees are motivated and dedicated, it is also important that they are provided with adequate tools which enable them to speak their own natural language of technology, the language that makes them indeed digital natives. Tools can be in the form of hardware or software and this chapter therefore aims to go through the design process of tools that would enable digital natives to integrate in the development process of new tools.

© Springer-Verlag Berlin Heidelberg 2015
A. Dingli and D. Seychell, *The New Digital Natives*,
DOI: 10.1007/978-3-662-46590-5_6

6.1   Choosing the Process

A process framework is a set of activities that are applicable for software projects [258]. This section introduces the key activities of these processes while assessing how a process can be effectively used to create software for digital natives.

6.1.1   Design Process

Before looking at the actual models and possibilities, one must first understand and accept the limitations at hand. The two main actors in the process are the second generation of digital natives who will be the users of this software and the computers or devices that would enable the dissemination and use of the designated applications. It is understandable that children and young people would find it difficult to conform to what we have traditionally perceived as a development process. In these processes, one generally assumes that users are included by applying methods such as meetings and focus groups. While these can still be considered, one has to to think further outside these processes and consider new methods of including the 2DNs.

2DNs will fit in these methodologies in a similar way as previous generations. However, as discussed in the previous chapters, they would only be maximising their effectiveness and contribution if they are provided with the right environment and tools. In an effort to consolidate existing research and practice in the area, we propose the amalgamation of different process models to maximise the contribution of these users in the development of software that is ultimately targeted for their use. Two processes were considered to be the easiest to be adopted in order to facilitate the inclusion of the 2DNs. The first process is the Interaction Design Process [87] due to its focus on the understanding of user needs and its subsequent evaluation of how these needs are met by the designed software. The second process is the User-Centred Design (UCD) [269] which reaffirms the importance of including the user in the three major phases of Design, Implementation and Evaluation. UCD can be generally defined as the process in which users are consulted throughout, generally as evaluators [262] of each stage.

Users are not explicitly featured in Figure 6.1 but it makes sense to conceptually imagine the user right in the middle with the process taking place around him/her. Another interesting aspect is that it does not really have a start or an end. However, we stands to reason that the entry point is the design which is then developed into an implementation of itself (that is ultimately evaluated). The circular path without an exit signifies that the process does not have any set number of iterations and would take place as much as it is needed. This can be compared to the metaphor of polishing a rough material into a smooth and shiny state. With patience, the material has to be polished over and over again, applying the correct pressure according to the evolving state until it is ultimately smooth and shiny. However, the latter adjectives are not conclusive in themselves since one can keep polishing. Software shares this situation since it can always be refined but there is always a point where the initial objectives are satisfactorily met and the product can be considered as concluded.

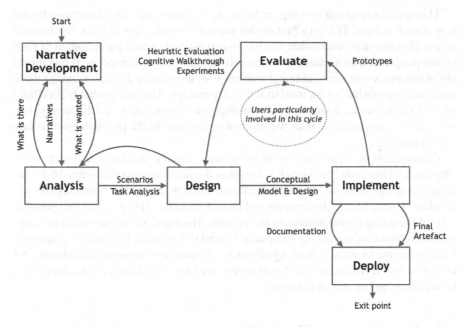

Fig. 6.1 A combine process to facilitate interaction design for the new digital natives (adapted from [87] and [269])

Figure 6.1 shows this simple approach that starts by developing narratives with digital natives. This process should iterate with proper analysis of the system and this would ensure that the scope of the application matches the need of the digital natives. Once the analysis is concluded, the design for initial prototypes takes place and is implemented in order to be evaluated. It is important that users are particularly involved in this cycle by holding scheduled and regular sessions that may take the form of focus groups to receive feedback about the prototypes and ascertain that it is inline with the initial objectives. After various design-implement-evaluate cycles the artefact is ready for deployment. The final artefact has to be accompanied by proper documentation that should ideally include summarised instructions that appeal to the new digital natives as discussed in earlier chapters.

6.1.2 Participatory Design

Those who were ever involved in development know well that whatever happens in the process is nothing more than a struggle between two major forces. These are the technical team and the user, who might also be the client. While this description might seem more of a dramatic narration, it is important that when considering the development of software for Digital Natives one also visualises the struggle that might not really be as tangible as in other situations. Participatory design is a sub-set of User Centred Design and it includes situations where there are different parties involved in the collaboration [262].

This is a new field that is being explored. Applications are being developed for the new digital natives. The very first digital natives are right now in their first teenage years. This therefore means that one has to first think about the participation of such young people, and even children, in the software development process. If one thinks about this process in a conventional manner, it would be very difficult to image these users abiding to what we are used to at this day and age. This will probably shed light on why participatory design might be problematic at times particularly because one needs to be knowledgable with the limitations inherent in the psychology of young digital natives.

Participation of digital natives in the process remains the key solution to these challenges. Hourcade [147] briefly discusses the involvement of children in the interaction design process. He facilitate classification by giving the following roles to children: Users, Testers, Informants and Design Partners. This is in fact one possible way of including digital natives in the process. Hourcade's roles are extremely important to make a case for the reasoning behind this chapter. However, we strongly believe that the ultimate role of digital natives is the role of continuous evaluator. An effective design process is one that evaluates prototypes and deliverables in relation to the needs and targets of the user.

6.2   Requirements of Digital Natives

"What do users need or want?" This is always an enigmatic question at the start of the design question and is very difficult to answer. The first challenge is that the conceptual model in the mind of the user is naturally not close to that in the mind of the designer. Furthermore, it also gets more complicated since users themselves are not always aware of what features and functionality they really need to find in their new software. This chapter aims to investigate this aspect while it explores why this question is even more difficult when one is dealing with children. Subsequently, we will propose techniques and adaptations of how users can be better involved in order to get the right answers from the early stages of the process.

6.2.1   What Do Digital Natives Want?

Before understanding how to do it, one should get to know what Digital Natives actually want. It is important to understand how we deal with adults when they themselves commission the development of software. This section therefore starts with a short scenario featuring an adult commissioning a system for his enterprise and then we shall explore how we ensure that such a process becomes effective with a younger audience.

Consider a fictitious non-technical entrepreneur running a small number grocery shops in a small town requesting an Enterprise Resource Planning (ERP) software for his business. He would have a vague idea why he needs such software to improve the operation of his business. He might also have a couple of features in mind that would facilitate his day to day running of this enterprise. However, when it comes to less tangible features such as data-mining, just-in-time ordering and similar concepts, such non-technical users might feel at a loss and would need to be supported with proper guidance. At this point, one must understand that the user would need to be educated about the possibilities and their respective benefits which might not be visible at first sight. Furthermore, the requirements engineers or designers do not need to simply educate the user. They also need to understand how operations are carried out within the cultural environment of this enterprise. This means that even when dealing with adults, one cannot expect they know what they really need from an application or have the linguistic tools to explain what they do and how such software can help.

In practice, this process is not really straight forward. This stage of the life cycle is also generally underestimated by both development teams and users alike. Whenever this is the case, the project ends up being derailed from the initial plan and budget. It therefore requires a highly disciplined approach and one must add that patience is also very important for both sides.

When developing applications for the 2DNs, it is very important that one gets into certain depth at this stage of the process. This stage needs to be approached with caution and better understanding of the user profile. The following set of issues and questions are intended to help the reader realise certain limitations inherent in the process of understanding what young users need and how they can contribute to the actual requirements gathering stage:

1) Can they visualise that what they are doing now would result in some app?
2) Do they have the linguistic tools to express themselves about what they are thinking?
3) Do they really care at this point about the application that one is trying to design?
4) Can they be engaged for a sufficient amount of time to properly provide the designer with an idea of what they need?
5) They would not be the ones commissioning the project and thus won't be directly motivated by milestones and progress of development.

All these issues point to a single conclusion. The possible reluctance of digital natives to properly answer questions related to what they are thinking, doing and needing reduces the choice of techniques at hand to elicit their needs. Designers therefore have to apply qualitative field research methods to better understand digital natives. These techniques are normally classified into two categories, Indirect and Direct methods.

6.2.1.1 Eliciting Qualitative Data Indirectly

Indirect methods include Diary Analysis and Interaction Logging. Diary Analysis when designers study the diary entries of the users in relation to a process that they are currently undertaking [87]. This can be a process that directly relates to an application or any other process that would lead to the development of new software. In the case of digital natives, it might be difficult to have subjects keeping a diary.

Even if they do, it would be difficult to interpret due to the possible weak vocabulary of the subject (in case of small children) or the lack of understanding of the entire process. It is therefore recommended that in this case, someone who is understanding the cause keeps a short diary in the name of the digital native in order to assess better the needs of the subject. Diaries are very important since they shed light on routines and while giving a temporal meaning to the way processes are done. On the other hand, Interaction Logging can take the form of an automated process which can be very easily used with digital natives. This process involves the recording of actions carried out by the user when using the software that vary from clicks, movements, tapping and usage of buttons. By the end of the process, the designer could then map the interaction log with the actual task carried out. This correlation would then identify any possible difficulties in the process and what was being done when a difficulty was encountered.

In the past years we have seen the increase in the use of tablets by children or even toddlers. Tablet computers are very convenient for parents to let their children draw or paint while not making a physical mess out of the activity. Furthermore, tablets also offer a variety of applications that children can use to entertain and/or to educate themselves (processes such as learning the number systems by playing different games). We have all possibly witnessed toddlers who are not yet capable of speaking but can very easily navigate a tablet computer. The toddler would be drawing on a paint application then, all by him/herself, pressing the home button, gently swipe through the screen of icons and choosing the next application that could for instance be a game. This simple narration given here could be the account of someone supervising the child using the device, such as for example parents keeping a usage diary for diary analysis in the design process. Simultaneously, there could be a background application running on the device recording the actions of the child on the tablet. The narration on its own does not indicate how long the child took to press the home screen after he/she stopped using the paint software. It does not even say how long the child took to start swiping through the icon pages and how long it took the child to actually choose an icon. Therefore, a balance of these two methods would help in the indirect elicitation of qualitative data. It is sometimes suggested [10] that personas are used when children are not available. However, this might cause serious difficulties when designing for digital natives since the person role-playing the persona or actually writing the profile could be biased towards the realities of a digital immigrant. Furthermore, the persona would be the digital native through the lens of a digital immigrant that would be more out of focus when trying to understand the realities of 2DNs.

6.2.1.2   Eliciting Qualitative Data through Direct Observation

The easiest way to think about this class of methods is that of understanding they all revolve around the observation of the user carrying out a task or a process. Digital natives might not expressively know how to explain their interest or preferences and this calls in for the need of carrying out observations without asking them much about it. Furthermore, digital immigrants also normally tend to seek a certain degree of comparison between the way they did things against the way digital natives do them. This comparison is helpful when it comes to redesign existing processes in order to allow digital natives to become more efficient at carrying out the designated task.

Ethnography is a way of observing users carrying out a task without interrupting them. While ethnography in interaction design is normally used to understand the social characteristics of the environment where the task is being performed, it can be used to study how younger people are using an application. Ethnography aims to uncover certain skills that might not be evident at first sight but which are key to the completion of the task or process. This approach is normally carried out at requirements stage in order to understand what users need and their respective interest in using the application in relation to a process. Nevertheless, it can also be very useful at other stages of the design process in a continuous effort to evaluate the deliverables by adding a flavour of heuristics that are discussed later in this chapter. Druin [92] highlighted the importance of having children as informants in the software process. The way digital natives to use software and the way they think about it might not be the same thing that digital immigrants perceive. Therefore, their role as informants must be strengthened and flourished in the entire process to motivate them to contribute as much as they can to the actual development of the application. This might be generally known as Protocol Analysis or Think-aloud Protocols. The challenge outlined above is that unlike older users, children might find it difficult to explain what they are doing and why they took certain decisions while using the application. The designer should then facilitate the process by asking adequate and timely questions to these young users in order to get the same effect of protocol analysis.

6.2.2 What Do Digital Natives Do?

This question cannot be answered generically. The answer is application specific and the challenge in this stage of the development process is to tailor an approach that can facilitate the elicitation of an answer from digital natives that is tailored for the application in question while properly following a common structure.

Task Analysis is the set of steps taken to understand how and what people do to perform a task or a job [87]. This varies from the things one does to perform a task to the actual tools being used to carry out the designated task. Needless to say that there is also a degree of knowledge inherent in users carrying out a task that might be taken for granted.

Whenever there is the development process of an application, designers and developers need to clarify the set of steps being taken by the user to carry out the tasks required by the said application. In Chapter 4 we provided two examples of different classes of applications that are widely used by digital natives, namely games and productivity applications. Since games have their own particular properties related to design and development, this example focuses on productivity applications.

Consider a fictitious application 'myView' that allows different users to enrol together in a group, in terms of their physical proximity and then film a scene with their mobile devices. Each user would have his own view of the same event and then, their devices would communicate and aggregate the content in a single video. This can then be either converted to a 3D scene or simply edited and uploaded instantly to a video sharing website such as YouTube.

In order to properly study the task, one needs to undertake these 3 main processes. The first one is **Task decomoposition** where the task at hand is split into sub-tasks [87] and ordered in a way that would combine into the original task. On the other hand, **Knowledge-based techniques** strongly consider what the user needs to know in order to carry out the task [87]. This also includes the actions involved and the knowledge inherent in these actions. From the designer's perspective, it is also important to know the relationship between the user and the application. This is then modelled using the **Entity-relation-based analysis** that is an object based approach where actors and objects are linked [87]. The above example of the myView app would need to be properly decomposed in order to plan development in a more efficient manner. On the other hand, the end result might not be so clear for the user at first sight and therefore an overall view of the system needs to be given for everyone to understand the link between the users, devices and tasks. Ultimately, one must also conclude which skills are needed by the users to set up this application and properly get the devices working together in order to achieve the ultimate desired result. Applying all this to the myView example would trigger various thoughts that might be overlooked on the first time round.

6.3   Evaluating Applications with Digital Natives

Evaluation is the process of ensuring that the deliverable meets the required criteria established at requirements level [87]. It is also agreed that evaluation should not be a single step that happens once in the design process just like it used to be traditionally depicted in the waterfall model. Constant and consistent evaluation is very important to ensure that the scope of the entire application is preserved. This means that everything has to be evaluated right from the initial steps in order to ensure that what is being specified is ultimately the real need of the user and this pace has to be maintained up till the very end of the process. Subsequently, evaluation is needed to ascertain that the design reflects the designated requirements of the user within his/her context. Finally, the deliverables resulting from the implementation of designs have to also be evaluated to ultimately verify that these deliverables reflect the needs of the user that were meant to use them in the first place.

Dix [87] explains that evaluation has three main aims. The first is to ascertain that the required functionality is covered by the final product. Secondly, evaluation must ensure that the product is ultimately usable and therefore having a positive user experience. Throughout the process, the ultimate aim of evaluation techniques is to then ensure that any problems present in the software are identified and subsequently mitigated or ideally, eliminated. When evaluating for digital natives, the aims of evaluation remain practically the same, however, the first two aims have to be put into different perspectives. While the third aim of ensuring that problems are identified and resolved remains relevant in any context, functionality and user experience may be somewhat subjective when dealing with software design for digital natives. As explored earlier in this chapter, digital natives do not always know what they need or do not have the language to explain such needs. This is naturally carried forward to the evaluation stage where they also lack the language to explain what is wrong with the system. Therefore, both coverage of functionality and user experience become more challenging to measure. Throughout our experience, practice showed that user stories should be used to reflect the needs of users right from the early stages of development and these are to be kept present and referenced till the very end. During the evaluation process, user stories are then used to verify that the initial plans were followed and delivered. Digital natives have then to be observed using the app itself and one has to be careful to notice the degree of ease that subjects undertake when performing tasks on the software in question. Gesture tracking software can also be used to measure the actions of digital natives and these can also indicate the degree of effort one needed to carry out a task.

6.3.1 Heuristic Evaluation

This commonly used method of evaluation is guided by a set of rules that loosely indicates what needs to be noticed and catered for. This is generally linked to the evaluation of usability but there are other uses for this evaluation. Some of these heuristics include:

Visibility of the System

Users need to be updated about the status of the system. It is not enough to simply provide movement on the screen when the machine is working. Visibility of the system refers to informative feedback that is helpful to the user [196]. An example of good visibility would be a progress bar outlining the percentage of the loading time together with a possible indication of the time until the process finishes. This can also apply to more detailed tasks such as uploading a file while illustrating the percentage of the file that has been uploaded till then.

Match between system and real world

The system has to present any form of information in a way which users understand. At a simplistic level, this heuristic refers to metaphors adapted in the system in order to link functionality to real-life concepts that are already familiar to the user. A practical example would be an icon of a musical note to refer to music. This even goes beyond the general understanding of functionality.

User Control

Users expect to be in control of the software they are using. It is therefore imperative that users are provided with paths of use in the software that would enable them to back track and freely move around the application [87]. The provision of undo and redo functionality in applications is one of the key approaches that provides users with the required degree of freedom.

Consistency and standards

New applications should follow a certain trend of consistency with respect to other applications with which users are familiar [196]. It is accepted and understood that particular common words such as 'file', 'edit' and 'help' are used throughout in different applications, carrying the very same meaning and implications throughout. This prevents the user having to learn new applications and platforms afresh.

Error Prevention

All users are prone to errors when using a system. In an effort to minimise errors, the requirements stage should be carried out as closely as possible to the end user in order to reflect the real way a user carries out tasks on a system. As explained above, this might be relatively easier when the users can actually explain what they need to do. However, when tailoring software for digital natives, one must bear in mind that they are not really familiar with the tasks they are to carry out and therefore, it is more challenging to design a system in a way that it can prevent errors (which are user is prone to make) while using the application in question. A traditional example of this heuristic was the warning, given by the system if the caps lock was on while the user was about to enter a password. For younger users, it is best to limit the possible options in an effort to reduce errors that users might not be prepared to handle while at the same time, allowing a certain degree of freedom in the use of the app. On the other hand, there are other emerging situations in the modern use of technology that need to be properly handled in order to avoid certain errors. The increasing dependance on the cloud is one of them. As systems become more modular and distributed, users must be adequately warned about such dependancies in order to prevent a single module to ruin or interrupt another process. Underlaying model checking techniques should be therefore employed to handle the emerging complexity while guaranteeing a smooth experience to the user of the system.

Recognition rather than recall

Flow tends to be smoother when the mental effort on the user is low. The use of software is a sequence of decision that need to be taken as fast as possible and the more time a user takes to carry out each and every decision, the more tedious the use of the software is. This means that users should not be required to remember any information in order to move from one screen of the application to the next, or backwards. If information is carefully depicted and organise, the flow of software can feel natural, ideally matching real life situations as mentioned in the second heuristic. This is what ultimately allows us to easily switch from one application to the next while softening the learning curve of new software at hand.

Flexibility of use

The goal behind the use of software is to carry out a task. The application has to provide various ways in which the user can get through an app in order to achieve a task. The traditional example is the use of keyboard shortcuts as an alternative to functionality available through the user interface. These are also known as 'accelerators' [196]. For example, to copy and paste, a user may opt to use the 'edit' menu, right clicking, using a toolbar icon or using the keyboard shortcuts. In this way, the design would be able to cater for advanced users who can carry out the task using keyboard shortcuts to novice users navigating through menus. This is also very relevant for software that is designed for digital natives who might, in their own way, develop their own ways of navigating through it. Designers should not try to predict the path taken by the users but alternatively, provide users with a variety of paths which enable them to create their own way through the task.

Minimalistic Design

The user needs to be presented with the right amount of information and therefore, the design has to be minimalistic in nature. Content has to be presented in a simple and clear manner. Buttons or other user interface component that are linked to functionality that is not specifically needed on a screen, should also be removed from the interface. This leaves the user with access to functionality that is ultimately needed to carry out the task at hand.

Recovery from errors

It is expected that users encounter errors while using any applications. It is therefore very important to acknowledge this and design software accordingly. First of all, errors should be explained in a clear and easy manner for users to understand thus avoiding error codes or other complex messages on the user screen [228]. Younger users might have a poor conceptualisation skills and therefore error messages must be straight to the point and with clear guidance on how to recover from these errors. Systems should handle errors independently from the user and the process of identification of an error together with its recovery should be hidden as much as possible from the user. Errors are confusing for inexperienced users and therefore systems should be built in such a way as to communicate errors to the users depending on his/her level of expertise.

Help and Documentation

Support documents need to be provided to the user in case of difficulties. Providing help and documentation on how the system works is important to support the idea of smooth flow in software. There are different ideas about how to go about this heuristic but it ultimately boils down to how information is provided to the end user. This can take the form of the traditional text messages. On a note inspired by game tutorials, information can also be provided in multimedia, hence images, audio and even videos if needs be. The strategy of when to display such information is also crucial. This can be accessible all the time or provided just in time.

This method of using usability heuristics was tried with children [196] and it was found to work if the said heuristics were adequately explained to the participants. It was however outlined that the wording needs to be rephrased in order to ensure a more effective approach. From the experiments that we carried out, it is also recommended that children are not only guided through the heuristics but given actual examples of each heuristic in order to link them to what they already know. When using graphical examples, it is also safer since other barriers such as language are mitigated and this renders the entire evaluation process more effective.

6.4   Its All about the User

This chapter aims to highlight the software design processes that are more adequate
to be used with new digital natives. User involvement is undisputed, however, this
may bring around various complications when dealing with digital natives. It is very
important for designers to be well aware of the limitations that digital natives have
to express when contributing to the software design. The requirements process has
to be as friendly as possible with significant interaction so that users would remain
engaged in the process in an effort to receive as much information as possible from
them. The findings during the requirements stage should then be properly translated
into adequate prototypes that would motivate digital natives to give better feedback
in a process that has to iterate on design, implementation and its evaluation. Evaluation is key in modern design techniques and in this chapter, we recommend the
use of heuristic evaluation [228] with particular adaptation to digital natives. This
technique remains relevant and the heuristics must act as a checklist for designers
to make sure that the application is as friendly as possible.

Chapter 7
Smart Homes

[257] defines a smart home as being

> "a home equipped with lighting, heating, and electronic devices that can be controlled remotely by smartphone or computer: you can contact your smart home on the Internet to make sure the dinner is cooked, the central heating is on, the curtains are drawn, and a gas fire is roaring in the grate when you get home".

According to [135], this term was first officially used in 1984 by the American Association of House of builders when they created a special interest group called 'Smart House', whose aim was to push forward the inclusion of the necessary technology into the design of new homes. This was a natural progression considering the massive advancements in technology which occurred throughout the 20th Century. In the early 1900s, it was mainly about introducing vacuum cleaners and sewing machines in the homes. This occurred mainly due to labour shortages, however technology spread like wildfire and in the following years, it became common amongst most households. Aldrich in [135] writes that By 1940 the proportion of households in UK with mains electricity had risen to around 65 per cent.

After the Second World War, kitchen appliances (such as electric cookers and refrigerators) became widespread soon followed by the introduction of televisions. This lead to the dependence on electricity to increase drastically and so did the consumption. In fact, during the 60s and the 70s, small electric devices became common in the homes. Also, these devices made the transition from being a luxury device towards becoming a necessity in the home.

The 80s and the 90s saw the boom of commodity electronics such as televisions, video recorders, DVD players, cordless mobile phones, etc. This was also the period when the notion of a home personal computer (PC) was conceived and sold to the masses. Soon after, these PCs were connected to the internet thus opening a myriad of possibilities for home users.

© Springer-Verlag Berlin Heidelberg 2015
A. Dingli and D. Seychell, *The New Digital Natives*,
DOI: 10.1007/978-3-662-46590-5_7

7.1 Smart Homes Today

The concept of a smart home evolved radically in the past decades with the proliferation of the internet whereby the idea of a house full of interconnected devices
started to take shape. Home automation is estimated to rise drastically in the coming years reaching 90 million homes by 2017 according to [263]. However, the real
breakthrough will happen when machines will start communicating with each other
seamlessly and using open protocols. This milestone is so important that 2012 was
seen by many ([322], [278], etc.) as being the year of Machine-to-Machine (M2M)
communication. It was also estimated that the number of devices connected would
rise from 1 billion to 2010 to 50 billion in 2020. This would ensure that cheap,
small devices can communicate with each other using technologies such as radio. It
would also give rise to a mesh of self-assembling networks. They would be aware
of their position having GPS capabilities. Thus, imagine street lighting enabled with
these capabilities. Maybe they could switch off if there's no-one around thus saving
electricity. In case of malfunction, they can report it directly thus avoiding having
people checking over them. The possibilities are practically endless.

Apart from providing us with a new level of commodity whereby devices are
intelligent enough to adapt themselves to our needs, there is also another important
use which is driving home automation to new levels. This is normally referred to as
Ambient Assisted Living (AAL) which is the idea of using technology in order to
prolong the time elderly or ill people stay at home in a secure and safe automated environment. This is being done through various means such as programmable washing machines, cookers and other home appliances capable of adjusting themselves
in advance in order to suite the comfort of the user. In order to ensure the well-being
of the elderly or the sick; devices exist that signal emergencies when the user is in
distress. Further still, projects such as [79], [77] and [78] ensure that these technologies make use of off-the-shelf devices thus keeping costs incredibly low.

The current state of smart homes is far from ideal, but as time goes by, we are
seeing a quick advancement in modern technologies. There are quite a couple of
challenges that need looking into if we want it to become a reality. One of these
is the issue of reliability and efficiency in sensory systems. Another major factor
is that of standardisation of information and communication systems. We will dive
further into some of these challenges in the coming sections.

7.2 Home Automation

There are various definitions for home automation due to the fact that the term could
mean very different things for different people. In general, a very vague definition
can be anything that can give you remote control of any home appliance or device.
This could range from the control of lights to a very complex security system which
uses natural interfaces such as voice control. These are just two examples out of the
many in the subset of home automation.

Home automation has various benefits. The first of these tackles the energy-saving problem. Through home automation, it is now possible to entrust your smart home to regulate appliances to work in the most energy efficient manner whilst taking advantages of low energy tariffs (such as those adopted during the night). This benefit is obviously a win-win situation, where the consumer will take advantage from a reduction in his expenses and the environment is benefiting from the reduction in carbon emissions because less energy is being utilised. There are various ways in which this can be achieved such as ensuring that devices (E.g. light switches) are off when people are not in their vicinity. Mobile devices can be used to connect with these electronic devices via the internet, check their status and turn them off from anywhere in the world. Another way of using intelligent home automation systems is by installing sensors and setting the actuators to adjust according to one's need thus ensuring that no electricity is used apart from what is absolutely necessary [178]. We can see this done with a large variation of devices. One example of this is when a light sensor is attached to a dimmer. The dimmer will set the brightness of the lights according to the amount of natural light detected. Since the sensors are usually battery operated, it is fundamental to have an energy efficient routing system. [230] proposes a scheme called RDSR (Relative Direction based Sensor Routing), whereby a home is divided into sectors and one manager node is placed in each sector. This node is responsible for passing all the data from the sensors to the base station through the shortest path on an 2-Dimensional plane, which in response sends the respective messages to the controllers.

Another clear benefit of home automation is the notion of convenience. This is due to the fact most device can be controlled from a remote location. It means that devices could be turned on or off using a mobile phone to suite the user's commodity. This can also help to reassure users that their domestic devices are off without the need of heading back to check. Furthermore, another benefit is the issue of security. High security home systems could be implemented with modern technologies. These can function with various features such as voice recognition or finger print recognition, amongst others. Also, live feeds of the security cameras can also be accessed through the mobile phone thus giving the user a 24/7 remote connection to the home. Home automation could also ensure that the house takes immediate action if there is a break-in. All lights could go on and multiple alarm could sound in order to scare away the intruder. This can be in addition to a warning sent to the residents mobile and to the local authorities.

7.2.1 The Technologies

To have such a system in place, various forms of networks are needed for the devices to communicate with each other as well as for the user to communicate with these devices. These networks can be classified into three different groups:

1. Devices interconnected using standard technologies such as Bluetooth[1], Firewire[2], USB[3] and others. Wired networks such as USB and Firewire are becoming less common in home automation due to the fact that people are opting for wireless networks.
2. Specialised control and automation networks, which consist of various types of networks such as X10[4], BACnet[5], EHS[6], ZigBee[7] amongst many others.
3. Devices capable of accessing data networks such as Ethernet, WiFi and more. These networks are very common, inexpensive and fast. They can be both wired and wireless. In particular, WiFi is becoming very popular since it is flexible, scalable and can be accessed by most mobile devices. One obvious disadvantage is that it tends to be slower than cabled networks.

Apart from the networking, one also needs the devices which will communicate over these networks. These can be either controllers, actuators or sensors. Controllers are those which process the information and decide on the action to be taken (if any). They are either hardware or software based. The programmable controllers have many advantages over a home PC, since they are less prone to crashing or suffer from security breaches thus making them more reliable. Actuators are those devices which act on the environment thus modifying it. The can range from complex machines such as robotic arms to simple light switches. Sensors are those devices which gather information from the environment such as microphones, cameras, thermometers, etc. These sensors are able to monitor a range of conditions and in response, send messages to the actuators who would take immediate action. An example of this is a thermostat sensor which would send messages to the air condition which would then modify the temperature accordingly. These sensors do not necessarily need to be placed next to the air conditioner but they are located at strategic positions around the room and then, the information is sent to the air conditioner using wireless technology. The architectures which controls these devices are either based around a centralised controller or around multiple intelligent devices which are independent of each other.

---

[1] This form of wireless network can connect devices within a 100m radius with a speed of 1Mbps [296]. The fact that it is a relatively cheap form of connecting devices makes it one of the more popular ways of interconnecting devices.

[2] The IEEE 1394, also known as Firewall, is a high speed, inexpensive way of connecting digital devices. According to [140], its scalable architecture and flexible peer-to-peer topology make 1394 ideal for connecting devices from computers and hard drives, to digital audio and video devices.

[3] [1] states that a USB is a serial communication link capable of speeds up to 4.8 Gigabits per second. USB protocols are capable of configurating devices both at startup or during runtime.

[4] A communication protocol which piggybacks on existing power lines found in the home.

[5] A Data Communication Protocol for Building Automation and Control Networks.

[6] European Home System.

[7] A personal area network using a suite of high level communication protocols which uses small, low-power digital radios.

7.2.2 Smart Interfaces

One of the most fundamental aspects of a home automation system is the way the user interacts with the system. The most common these days, is an internet based application on a mobile or tablet device. Designing a user interface is not easy when one has multiple devices since it is common to create a separate controller for each device. Even though this is probably the least complicated way of implementing it, it will create quite a few limitations. One such limitation is the lack of ability to set a particular ambience at the touch of a button since all the different devices have to be set individually. Most of these devices also lack internet connectivity thus making it difficult for the use to control them via the web.

An integrated solution is normally much better but it normally limits the user to the products of a particular brand. Having different devices controlled by the same protocol gives the user full access to each device individually as well as different mode settings. Normally they are controlled either through their integrated controller or through a web interface. These mode settings could set a number of devices to a particular mode at the same time depending on the situation or particular occasion. Also, one section of the application could be designed in a way as to control all the devices in a particular room or area of the house. Information about the temperature in each room can also be stored on this application together with a live feed from the cameras installed. This application can also keep a log of events that are scheduled to happen at a particular time and a log of all activity pertaining to the connected devices. One advantage of this type of interface is that it can be controlled from anywhere as long as one has access to a computer and an internet connection.

When these devices are controlled remotely, it is of the utmost importance that different mobiles are able to access the same system due to the fact that very often, the house has more than one resident. Obviously, different people living in the house might have different levels of access. This means that everyone has access to their own room but not to each others. This is a very practical approach but we can still go a step further by introducing an intelligent system.

Such a system can be implemented in the form of an avatar. This is a digital human being transmitted through various screens installed around the house. The avatar could detect who is located in which room of the house through motion sensors and it can adjust the ambience of the room based upon the requirements of that user. In this form of user interface, besides the motion sensors, input and output is normally controlled through microphones and speakers respectively, as well as through multiple monitors. One such avatar is the one developed for the companions project [86] [249] [327] [103] where the system easily dealt with cooking devices, audio visual devices, healthy lifestyle and generic conversations. In fact, most of the interaction happened using natural language which is easier to handle for the user than a complex graphical interface on a mobile device.

All these user interfaces have their own advantages and disadvantages but the best and most cost-effective user interface is usually a mixture of multiple forms. For example, a hybrid system, merging the previous avatar system with a mobile based application. This tends to eliminate most of the disadvantages mentioned earlier and it also offers what is probably the most ideal user interface.

7.2.3 Smart Devices

The most popular tasks related to home automation normally include automated adjustment of heating, ventilation, air conditioning, lighting, audio visual, gardening and security. In fact, it is normally referred to as HVAC (Heating, Ventilation and Air Conditioning) automation [301] [289] [40]. The ventilation could be automated by installing sensors to monitor the quality of the air. This in return would initiate the ventilator in order to change the air and turn it off once the right quality of air is reached. With regard to heating and air conditioning, a thermostat is introduced which would send data to the controllers thus adjusting the devices according to the temperature. Another way of automating the heating and cooling is by initiating the devices with timers or by using a mobile device. This is done to ensure a comfortable home, since the right temperature can be set while the person is on his way home. These systems are designed to provide complete comfort while also ensuring an energy-efficient environment.

Another type of system that could be easily implemented in home automation systems is a lighting system [122] [218]. This can be controlled in a number of different ways:

- A light sensor can be used to dim or brighten the lights automatically depending on the amount of natural light available in the room.
- Motion sensors can be attached to individual bulbs thus ensuring that if nobody is detected in a room, the lights will go off only to go on again when there is some sort of movement.
- Electronic timers ensures that lights are switched on and off at predefined times of the day.
- Lighting can be controlled using mobile devices which can easily contact the light bulb and adjust its brightness or even colour, depending on the mood of the user. Further still, intelligent lighting systems such as the one proposed in [128] are capable of learning and adapting themselves to the needs of the user.

With regard to Audio Visual devices, there are more features and configurations. The sound system, lighting system and the television can be connected together in order to set an ambience depending on the genre of the film being watched. Audio automation can also be installed around the house using different sensors, thus changing the ambience of the whole house. The sound will play on specific speakers depending on your location within the house and will keep on following you as you move along. In this case, these automation systems are installed for pure comfort and have no additional functionality such as energy saving.

A fundamental part of any automated home is its security system [268] [165] [74] [36]. By security, we do not just refer to the prevention of break-ins but also the prevention of fires, gas leaks, floods and other house hazards. A number of sensors, which monitor for specific signs, could be installed around the house. If there's an alert, a warning signal can be transmitted in a number of ways such as an SMS, e-mail, phone or through the traditional alarm sound. Security cameras can also transmit live feeds directly to a mobile phone so that the occupant can be given full access to his security cameras from anywhere in the world.

Besides these main features, found in home automation systems, several other devices can be easily automated.

- One such example is gardening. Automated watering of the plants [188] [9] is becoming more common by combining timers with sprinklers. Also, special sensors could be implemented into the soil which monitors the quality and state of the soil.
- Simple day-to-day tasks such as coffee making, switching the washing machine, etc can also be automated with a timer or with a smart phone since homes are being installed with wifi networks and more devices are being connected to the internet.
- A smart couch could also be installed which could monitor and give feedback on ones posture and possibly adjust the shape accordingly. It could also intensify the film watching experience by introducing the fourth dimension in the form of vibration.
- Sensors can be installed in letter boxes, capable of notifying the occupant once a mail is received.
- Smart picture frames could also be installed around the house. These could store important reminders that could alert the occupant at a particular time no matter where he is situated in the house.

Each room in the house can be easily turned into a smart room. This means that in every room, certain devices are installed which can provide the necessary aids which are usually associated with that room. In the next sections, we will be discussing a number of features that can be implemented in particular rooms of the house.

7.2.4 Smart Kitchens

There are many features which can be easily installed in the kitchen [13] [299] [235] [53] [56]. An appliance that could be implemented is a smart fridge. First and foremost, this can easily contain a touch-screen on the front which will act as the controller for all the kitchen appliances. This fridge can be equipped with a set of sensors on the fridge installed strategically. These sensors will monitor the objects and the quantity that is actually left. It will then automatically draw up a shopping lists which will be sent to a supermarket which offers online purchasing. Expiry dates will be automatically checked by the system thus preventing unpleasant surprises in the refrigerator. Another feature which this fridge might possibly have is a database of recipes. This database could compare the items contained in the fridge at that time together with the ingredients of the recipes and provide a list of different recipes to choose from. Furthermore, once a recipe has been chosen, the oven can be automatically turned on and set to the right setting automatically. It could also provide instructions on the method of cooking as the cooking progresses. This can be easily achieved by using NutriSmart, edible RFIDs [310] embedded directly in our food. The food tracking system will allow consumers to trace the entire supply chain of anything they consume. It would alert people with allergies or dieters about the ingredients used in the preparation of the product being consumed. Apart from this, [136] and [280] propose a system where different cameras are installed in the kitchen. These are placed above the sink, on top of the stove and above the counter. An algorithm is developed in order to monitor attributes associated with cooking such as methods of preparation and cooking techniques. It could be taken a step further where information regarding the nutrition of the particular meal is given using these cameras in combination with the edible RFIDs mentioned earlier.

7.2.5 Smart Bathrooms

Another room that could possibly be converted into a smart room is the bathroom [238] [201]. In this case, privacy becomes a major factor since filming inside the bathroom might be considered as rather invasive. However, a fully automated system such as the one proposed in [79] whereby there is no human monitoring of the footage but only algorithmic analysis of the film might be more acceptable for some people. Another possible solution is to use sensors hidden around the bathroom in order to detect the whereabouts of the person. They could be easily installed in the flushing, thus monitoring and calibrating appropriate water usage. Another possible feature in a smart bathroom could be the information displayed on the mirror [110] [145]. This can display all sorts of information ranging from weather, news, calendar, medical, etc as well as health and beauty tips. Since that the bathroom can also be a place of total relaxation, one can easily include automated speakers to play soothing songs and an automated light system. The bathtub can also be set to automatically fill up with water at a specified time and the toilet seat might be instructed to warm itself such as the one described in [311]. It is important to note that these systems help the user save energy without necessarily economising on the comforts of life. This is achieved because the various services (such as heating) are only used when they are needed thus there is no wastage of resources.

7.2.6 Smart Bedrooms

Whether it is sleeping, getting ready or just relaxing, the bedroom is the single room of the house where one usually spends most of his time. A number of smart devices could be installed in order to simply or improve the time spent in that room such as [182]. A smart wardrobe is probable every person's dream. The main feature of this wardrobe would be an integrated monitor such as [320] which displays different information (such as weather, appointments, etc) in order to help the person choose an adequate outfit. A database of ones clothes can be stored and displayed on this monitor. New collections can be downloaded during the night and displayed to the person whilst allowing them to shop directly from their wardrobe. One mini feature which this wardrobe might have is a weather system which displays the weather together with recommended outfits suitable for the current weather. It could also serve as an assistant suggesting clothes in case of travelling to different countries. Intelligent wakeup systems such as [189] coupled with light management or automated blinds could also enhance the dreaded waking up experience. This could be further enhanced by automating temperature control as well as preferred sounds for the wake-up call. A smart bed such as the one described in [138] could also be installed where it would monitor the user's sleep patterns in order to learn and adjust the ambience in the room according to his own preferences.

7.3 Ambient Assisted Living

AAL can be categorised into 3 sub-domains, namely emergency assistance, autonomy enhancement and comfort. Welfare technology is closely related to ambient assisted living. This is the technology that could help and assist users in their daily tasks. The main difference between these two is that whereas ambient assisted living mainly targets the elderly, welfare technology aims to assist different classes of people.

Welfare technology targets a number of different issues. It can provide support to different people with disabilities and give them a helping hand thus making them more independent. It also aims to help people with chronic illnesses live a life in the comfort of their own home rather than in an institutionalised home. Welfare technologies improve the quality of life of people by guiding them towards living a healthier lifestyle. One example of welfare technology is a bidet toilet which features integrated washing, height modification as well as others. This is ideal for those who suffer from mobility issues. Another way in which welfare technology can assist people undergoing some form of physical therapy is by integrating physical rehabilitation exercises in game consoles (such as [179], [192] and [55]) . Besides the fact that it is helping them perform their daily routine, it also includes an element of fun. Welfare technologies are also used to simplify the lives of people such as providing medicinal dispenser [297] with specific alarms in order to facilitate the intake of pills.

An issue which still needs to be sorted is the standardisation of technology due to the fact that the devices need to communicate between themselves if we want to gain maximum benefit from them. Also, it is very helpful if all automated devices in a home are controlled using one common interface. Standardisation can also help in reducing the costs of such devices since they will be based on common protocols. In the past years, we have seen a big drive from the Nordic countries (namely, Denmark, Sweden, Norway, Finland and Iceland) towards improving welfare technology. The countries of the European Union have invested billions of Euros through various organisations with this aim and in fact, the Nordic Centre for Welfare and Social Issues has setup a Nordic Innovation Network for welfare technologies which ensures that research and results are shared between various organisations in order to collectively improve the current standards of welfare technologies.

[161] clarifies that while Smart Living or Smart Home concepts mostly focus on Comfort, AAL puts Autonomy Enhancement and especially Emergency Assistance into the focus. Most of the automated devices that were mentioned earlier fall under the sub-domain of comfort. Even though they could be a great aid in the lives of the elderly, they are merely helping by providing extra comfort. This category is probably the easiest to implement because there exists all sorts of automated devices which are inexpensive and fall under this sub-domain. Although comfort living is fundamental, this should not be the main focus of Ambient Assisted Living.

Under the category of autonomy enhancement, we find a number of day-to-day fundamental tasks which if automated, could improve drastically, the standard of living of the elderly. These tasks could possibly cause a hazard for the elderly if not tackled in the correct way. A task which falls under this category is cleaning. Robots such as [107] [243] can be an incredible help for these people. Therefore, if tasks such as cleaning can be automated, the risk of injury by the elderly will drastically decrease. [163] stresses on the important of fall prevention and points out the with age, falls tend to increase. Autonomy enhancement surpasses comfort on the list of priorities, however it does not rank above the importance of emergency assistance. The vulnerability of the elderly is one of the main challenges in our society and it is expected to keep on increasing in the coming years. With systems that can automatically detect when there is something wrong and in that case, inform a relative, friend or even a carer (E.g. a hospital), quicker aid can be provided to the person in distress. A number of devices already exist to monitor these situations.

- The amalgamation of wearable or implantable multi sensory platforms with sensors set up in the environment that use low-power and dependable technologies, are being investigated and the first prototypes are being developed at the moment according to [315]. Obviously, using big, conspicuous wearable sensors or large implants can prove arduous for patients, so one of the major immediate challenges lies in the development of small and inconspicuous sensor systems that can, for instance, be embedded in clothes [195]. Low-bandwidth networks can be employed for the exchange of data on account of novel low-power wireless technologies.
- [14] are working on wireless networks of small sensors that monitor patients vital signs; such as cardiac activity. Martin Elixmann, the head of the group, states that people are often hospitalised all through their recovery solely because they need monitoring. The author explains that as the sensors are wireless, the wearers can move about the hospital or similarly in their home, freely. He asserts that existing devices on the market permit patients to be mobile; but they are merely able of record data, thus, nobody will become aware if the patient turns for the worse. On the other side, these sensor networks are capable of monitoring patients in real time and raise the alarm.
- [151] are taking a different approach whereby a software is being developed which examines images of the daily activities of a person. If an unexpected deviation from the pattern occurs, carers are immediately alerted.

- An intelligent Personal ECG Monitor (PEM), is being developed by [274]. Sophisticated decision-making techniques are embedded in the PEM, by means of which it is possible to produce diverse alarm levels and to send alarm messages to the appropriate care providers using wireless communication. The authors state that the care provider is drawn in only when essential, therefore making PEM a cost-saving solution. They affirm that PEM is a model of how healthcare can be improved via ubiquitous devices which are personalised and wearable.
- RFID can also prove useful especially when tracking prescriptions. They are vital for visually-impaired persons, where they can use devices located on prescription containers to verbally get information about the prescriptions and the dosage amounts, with the help of speech generation technologies [2]. HearMe is a medication management service idea, also targeted towards visually challenged older users, that converts medical information to speech [134]. The authors claim that this can assist visually impaired persons to distinguish medication and to retrieve information about its usage. They explain that this was achieved through the use of Near Field Communication (NFC) technologies, thereby enabling writing and reading of data from tags attached to medication boxes, after which, a speech synthesiser transforms the text to audio. In addition, RFID can be employed to help promote the self-medication of the elderly where patients would be able to track the location of their medicine and find out whether they have already taken it [162].
- AAL systems can also help in the rehabilitation of stroke survivors [308]. The tool offers users significant visual feedback of their own achievements, together with information regarding the quality of the movements and his progress against time. It is desirable to discharge patients from hospitals and support them in their own community with the provision of rehabilitation in their own home. They affirm that computer games and virtual reality systems are rising in popularity within the rehabilitation setting and are being seen as encouraging in endorsing exercise behaviour. On the other hand, the quality of movement, applied to finish a task in commercial games, can be frequently ignored, as the games tend to be more focused on concluding a task without focusing on the quality of the movement. Similarly, [302] sketchs an effective system for supporting physical activities, especially in home environments.
- [70] suggest a solution aspiring to create a home-based tele-care system supporting several chronic illnesses, for example; congestive heart failure, chronic pain and stroke, by utilising a variety of personal computers and healthcare technologies. They explain that the system is intended to endorse behaviour change through monitoring activity metrics and performing analysis, which in the end would produce feedback to the user in a significant and attractive way.
- Pogorelc in [253] recommends a health monitoring system whereby movement is captured with a motion capturing system and the output is modelled with a proposed time-series data mining method. He explains that the objective is to automatically distinguish falls together with health problems and to inform emergency services if required. Similarly, [326] proposes a system which

automatically detects falls using floor-mounted accelerometers to collect body-sound signals that characteristically happen in human falls.

- Context-detection algorithms together with fixed and wearable sensors can offer information that can activate messages at a suitable time. A history of the user in terms of physical activity recorded on it, can be developed to produce personalised feedback established upon earlier experiences and present context [315]. Smart Homes offers an infrastructure for putting into practice the concept of ageing in place, through a context-aware system that can be utilised pervasively [142].
- The most frequent chronic cardiovascular disease is congestive heart failure and as stated by the European Society of Cardiology (ESC), it is likely that in 2015, 12 million Europeans will have a heart failure [62] [172]. After discharge from hospital, some conditions, such as congestive heart failure, will need careful monitoring by doctors. However, this can face many obstacles; like being short of resources and facilities, patients not acting in conformity with the health care plan, etc which can lead to aggravation of the illness [264]. Tele-monitoring of congestive heart failure patients by monitoring and processing heart vibrations with a 1D or 3D accelerometer implanted under the skin, is another pertinent application that allows the detection of atrial and ventricle contractions, aortic and mitral flows, thus providing the hemodynamic status of the patient [315].

New studies have established that a patients remote home monitoring can improve the after discharge care which leads to more optimistic outcomes. Reynolds refers to a study by Sally C. Inglis [155], of Baker IDI Heart and Diabetes Institute in Melbourne Australia where 27 studies involving more than 9500 patients receiving heart failure management support, either due to telephone support or by way of tele-monitoring with digital, satellite, Bluetooth, broadband or wireless transmission of physiologic data were appraised. The patients condition calls for attentive monitoring of vital signs, (for example, blood pressure and heart rate, in addition to weight) as fast increase in weight can be a sign that the patient is holding fluid and as a result, a heart failure can occur. Both methods resulted in a fall in heart failure associated hospitalisations, as well as a better observation of the recommended treatment plan by the patients. Moreover, in nine of the studies, an important enhancement in the quality of life was apparent. Dr Inglis is quoted by Reynolds as stating that monitoring the condition of patients in their own home often, might potentially allow for the identification of the moment when the patients condition is declining.

Thousands of significantly ill patients are already relying on computerised health trackers to help keep them safe at home [109]. Freudenheim refers to a system being tested at present at the Mayo clinic. The article portrays a device that reminds the patient to monitor their health in the morning by lighting up and beeping. Patients are cued to put on a blood pressure cuff so that their blood and pulse readings are shown and to slip their forefinger into a sensor used to gauge blood oxygen. A scale linked to the device checks the patients weight. The device informs the patients if they are fine. It will exhibit a sequence of Yes/No questions as well and a nurse would get in touch with the patients if the answers are a cause of concern. Devices such as this can be of support in motivating patients, particularly mature ones who are suffering from chronic conditions, to abide by the doctors and nurses recommendations and to contribute to their own care. Doctors report that maintaining a constant health monitoring system is frequently less costly, turns out better health results than intermittent checkups and reduces recurring hospital stays.

As maintained by another study, heart failure patients using interactive tele-health systems with motivational support tools at home spent less time at hospital and their quality of life was described as considerably improved over an evaluation period lasting 12 months [224]. Patients were capable of taking vital measurements at their home and then convey the information to their doctors by means of the system. They were also equipped to obtain educational and motivational information from their doctors so as to assist them in managing their health. Monegain quoted Josep Lupon, the head of the Heart Failure Unit and chief researcher of the study as asserting that giving educational support to patients using their television, considerably contributed to their empowerment. In addition, Lupon asserts that the study demonstrated that passing around disease and patient specific information through their television improves the family members comprehension of how to support their loved ones and this gives them the impression of having a very strong influence on the results.

Predominantly with cardiovascular care, tele-medicine and home health monitoring is not something new, but this technology is taking on new dimensions [50]. Jonathan Edwards, research vice president for Gartner and a lead analyst on telemedicine is quoted as saying that, at the moment, the home monitoring area incorporates technology such as sensors for remote diagnosis, tele-retinal imaging, tele-radiology, remote cardiac monitoring, video conferencing, home and mobile health monitoring and counsel to patients. Edwards is quoted as saying that daily monitoring lets doctors and nurses keep an eye on their patients both in the event of an emergency and to prepare for any recent developments in their condition upfront. Furthermore, it can be likely that, someday, patients family will be allowed to access the information remotely, but it will hinge on the issues surrounding patient privacy. This would allow the incredible opportunity for families to care for their loved ones remotely because they would be capable of checking, for example, that their vital signs are in order, or that they are taking their medications or even monitor them through camera.

Such systems present the respite and independence of staying at home on top of the safety of being monitored daily and being given appropriate medical attention. Mainstream existing systems entail a compact monitor and a scale. The system accrues several vital signs, for instance, oxygen saturation, temperature, body weight, blood pressure, and heart rate every day. The data would be transmitted to a hospital where medical staff can evaluate the information and proceed appropriately. Should there be a variation in a patients health status, the medical staff will make contact with the patient. Benefits from such monitoring comprise proactive and preventative care; decrease in emergency room visits together with unforeseen hospital stays; perking up patient conformity while at the same time promoting patient education and self-management; early detection of clinical needs, alteration of the treatment plan and medication.

Hanak et al portray a mobile AAL solution which is designed to meet the needs of modern health services in caring for, monitoring and motivating the elderly in their own home. The authors suggest a solution that goes further than traditional tele-monitoring they take into account health management, mood assessment, mental monitoring, in addition to physical and relaxation exercise [132]. The authors explain that communication and delivery services are provided via built in GPS, Wi-Fi and 3G mobile connectivity and Bluetooth is used for blood pressure and body weight measurements through a body-mounted physiological sensor that monitors activity, stress and body temperature. They maintain that telemetric data is constantly recorded on a local host computer but also sent, at the same time, to a central database where a rule-based system makes emergency assessment and can also be forwarded for monitoring by health personnel.

Magjarevic has investigated early wheezing recognition [198]. The author explains that wheezing is frequently found in pulmonary pathologies and its detection is thought to be of significance for the diagnosis and management of respiratory diseases. Thus, the author developed a straightforward and robust system for its detection in respiratory sound spectra, which system is intended for long-term monitoring and early stage appraisal of asthma episodes. The author maintains that the algorithm is robust enough to be able to detect wheezing even in the presence of noise and moving artefacts and it is established on the concept of frequency domain peak detection.

A number of projects are currently being developed in diverse research teams. One of these proposed projects aspires to help with the disease management and medical care of chronic obstructive pulmonary disease (COPD). AMICAs (Autonomy Motivation and Individual Self-Management for COPD patients) major objective is to develop and weigh up long-term COPD management solutions established on new ICT that would permit for the early detection of COPD exacerbations; offers remote monitoring and home-based care; proffers a user-friendly design suitable for elderly users; promote prevention and self-management; and augment the levels of therapy fulfilment [117]. Foix asserts that on the whole, AMICA attempts to imitate the medical consultation at home by going through auscultation and an interview, which is realised by acquiring a string of physiological signals every day through ad-hoc sensors, which information is then extended by the responses provided by the patient when interacting with a dedicated mobile device. Foix continues to explain that by merging this information, the system is capable of setting off medical alarms, adapting minor parts of the patients treatment plan, lifestyle or even propose hospitalisation.

7.4   Challenges and Benefits

Even though we've seen a number of different systems, the use of several devices from varied vendors can bring about various problems, like:

- Uncommon standards for data exchange.
- Wireless and wired connectivity. Implementing a wired system would probably be more reliable than a wireless one. However, this may cause many restrictions as well as having the wires themselves being a potential hazard. On the other hand wireless connections are battery powered and therefore have a limited life which could cause many problems, especially if it is not noticed once the battery dies.
- Having no connectivity between the devices.
- Diverse models of usage, like using devices with the same function in another way [156].

[50] asserts that the chief obstacle is not the devices themselves but the transmittal of data from the devices and the lack in an infrastructure to monitor and identify the data. The information collected cannot be compared to normal health criteria without an appropriate recording system and consequently is of limited service to the doctors. Developing such recording system can be costly which in turn can prove problematic.

Additionally, a lot of obstacles that have to be overcome are not technical but related to business models, policy, reimbursement and cultural change in medical professions through new roles and redistribution of responsibilities. In fact, [315], suggest a roadmap for tele-monitoring and self-management of chronic diseases. The roadmap covers short-term, mid-term and long-term goals and tackles tele-monitoring of patient status transmitted parameters, wearable multi-sensor platforms, implantable multi-sensor platforms and self-management of chronic diseases. The short term goals, planned for 2013, focus more on the tele-monitoring of patient status transmitted parameters such as ECG, acceleration, weight, temperature, respiration, EEG, pressure, movements and heart sounds. This could be achieved through body worn or subcutaneous sensors. They suggest parameters like pH, pressure, blood flows, respiration, temperature and heart sounds can be monitored by the use of ingestible capsules. Sensors, whether body worn, ingested or subcutaneous are then connected wireless to cellular phones. 2014 actually saw the launch of a number of important initiatives in this area [219], primarily Apple's Health system and the proliferation of the fitness bracelets such as the Garmin Vivosmart, the Jawbone Up Move, the Microsoft Band and the Samsung Gear Fit.

The short term goal for wearable multi-sensor platforms is the provision of sensors whose power is supplied by rechargeable batteries and lead-less ECG and respiration monitoring. Power supply by batteries or rechargeable ones for sensors also feature as short term goals for implantable multi-sensor platforms. The short term goals for the self-management of chronic diseases include the involvement of the patients for powering and managing their chronic diseases.

The mid-term goals, foreseen as achievable by 2018, include external biomarkers for the tele-monitoring of patient status transmitted parameters. With regards to wearable multi-sensor platforms, the mid-term goals focus on powering wireless sensors by energy scavengers and this will also be used in the implantable multi-sensor platforms. The long-term goals, planned to be attained by 2025, fore-see implantable wireless biomarkers for the tele-monitoring of patient status trans-mitted parameters, wireless sensors supplied by energy scavengers, this time also for implantable multi-sensor platforms and they also predict fully-automated sys-tems whereby the patients need not intervene for the power and management of the chronic diseases.

When it comes to implementing AAL systems in the home, it is obviously much easier to install them during the construction of the house. However, the reality is that most houses are not constructed with this is mind. Therefore, the implementa-tion of an AAL system is not always trivial unless it is being installed in a new house which is specifically designed to suite such needs. Another possible challenge is the availability of services which can be too vast to deal with through robotic devices. Furthermore, these services are continuously changing and very subjective to the particular user.

Ultimately, the major challenge is the interface. Most elderly find it hard to adapt to complex interfaces and they would require one which is more natural, maybe through voice or gestures. If this is not provided, it would definitely make them hesitant to adopt such a system. On the other hand, some elderly citizens might not accept the fact that they are growing too old to keep on living the same life they've been accustomed to and therefore will be hesitant to rely on an automated system. The problem of lack of willingness or competence is one which will probably never be solved, even though it will decrease with time.

On the other hand, these systems will considerably improve the standard of living of these people. As we've seen already, a number of automated devices already exist in the prevention of certain injuries. However, it is currently impossible to prevent every injury or illness that is brought about by old age. Due to this fact, many devices exist and many more are in the pipeline to help where possible. Devices also exist which help people live with different disabilities and help them improve their standard of living. Whereas before, a helper would have to be employed to help with day-to-day tasks, nowadays, an automated systems can be easily installed. This would give the elderly the full control and satisfaction of living alone while being in full control of particular tasks.

Implementation of AAL systems can also raise a number of ethical issues. The first is related to privacy and the degree of intrusiveness in one's life. Obviously, one has to weight the pros and the cons in order to answer this question however, when a system is fully automated, one can argue that there are no privacy issues since no human has access to that information. Ultimately, it really depends on the individual. Some devices could be accepted by society while others could be considered as an invasion of privacy. One example of this problem is whether to use motion sensors to detect falls or a camera with an advanced algorithm. Even though they both exist, the camera is probably more reliable, whereas a motion sensor would be less of an invasion of privacy. The next issue concerns whether one can entrust the wellbeing of an individual within the hands of a computerised system. Even though these devices are thoroughly tested, the results are rarely 100% efficient. This means that although the devices would be very close to perfect, they could possibly raise false alarm. This could have an adverse effect on the individual and could ignite various trust issues. On the other hand, these devices might give a positive reading on the health of a person when the person would require urgent treatment. This could cause many complications with ones health.

Chapter 8
Digital Governance

There exist various definitions of Digital Governance. Some claim that it is a synonym for e-government, internet government, online government or connected government. [3] defines it as being the use of the Internet and the World Wide Web (WWW) in order to deliver government services to their citizens. [129] focuses on the fact that e-government should improve the efficiency and effectiveness of the various services. Whereas we agree with these definitions, we believe that they're incomplete and what we mean by digital governance is something wider.

Digital Governance is made up of two words. By digital, we are referring to anything related to the use of computer technology. The word Governance was derived from the Greek word and its metaphorical usage can be traced to [251] , a socratic dialogue written by Plato around 2,400 years ago. The meaning behind this verb is to steer or move in a particular direction. To do so, governments must take decisions which will impact on the administration and on the day to day lives of the citizens.

However even though this definition might sound pretty straight forward, it is our interpretation of the word digital that will give a new lease of life to this phrase. By digital, we are not only referring to digital technologies or the software which operates over that technology. We are not only referring to the presence of governments in the online world or the digital services which they offer to their citizens. But we are also referring to the digital lives of people and the issues related to governing those online citizens. We will explore topics like online citizenship, digital legislations, online politics, cyber crime and other issues related to this topic.

8.1 Online Citizenship

Citizenship according to [16] is the membership to a particular state within a territory, together with the rights and obligations attached to that status. This concept evolved through centuries of civilisations and in fact, we can trace, back in history, different forms of citizenships. If we peep back to 500 BC, the ruling class was an oligarchy[1] [4] thus ordinary citizens did not have much say in the decisions of their

---

[1] A small group of people having control of a state.

© Springer-Verlag Berlin Heidelberg 2015
A. Dingli and D. Seychell, *The New Digital Natives*,
DOI: 10.1007/978-3-662-46590-5_8

community. Eventually, the model became unsustainable and democracy (or the rule of the people) was adopted around the 5th century. However, democracy too had to evolve. For many years, the rule of the people consisted of adult men only, whilst women and children were excluded. In fact, [93] reports that women had to wait until the late 19th Century and early 20th Century for the right to vote in most western countries. Over the years, laws were created by citizens, they were then ordered and compiled into a constitution[2].

In recent years, the massive technological advancements like mobile devices, social media and e-mail changed the meaning of active citizenship. Citizens and their representatives can now maintain constant contact. However, we must be careful on the weight given to these channels. A study conducted by [223] claims that Twitter users are not representative of public opinion. The reasons behind this is because these users are normally younger, selective and because the reach of Twitter is quite modest in the online world. Researchers who use Facebook [217] [26] seem to reach similar conclusions albeit for different reasons. Using the current tools, we can only have a very rough estimate on the demographics of the people discussing particular topics in social media however, if we want exact results, we need adequate sampling without any bias. Having said that, we should not underestimate the power behind social media. We should not forget what happened during the Arab Spring[3] whereby Social Media played a major role according to [148] [304] [167] [150] [116]. From these analyses it transpired that;

- a lot of people where debating politics online
- these discussions eventually spiked before any major event on the ground
- it helped in spreading democratic ideologies and aspirations amongst the people

According to [260] Facebook played a pivotal role in the Egyptian protests. It gave people the opportunity to share information quickly through various mediums such as mobile devices. Apart from that, the system was not controlled by the regime, thus people could share their views and unite in order to overthrow the authoritarian regime. [167] claims that the new media is just one factor amongst the various social and political factors in these revolutions. However it played a critical role since in these countries, one can experience the lack of open media and the new media filled exactly this role. This was also brought about by the regimes themselves, as a side effect of globalisation, since such states could not halt the adoption and promotion of new media mainly due to economic compulsion.

The Internet's role is therefore unquestioned, however it is not the only factor as can be seen in [277]:

Neither Technology Nor Poverty Guarantees Revolution. There are other countries which are poorer and with a higher Internet penetration rate yet there is no revolution going on.

---

[2] The set of fundamental principles according to which a state is governed.

[3] The Arab Spring refers to the pro-democracy uprisings sweeping the Middle East and North Africa that began on the 18th December 2010.

The Medium Can Lead to the Message. Some of the bloggers taking part in the revolution actually started by expressing their grievances and this was later amplified by the internet crowd. A local message was blown into a global one.

Online Crowd Dynamics Mimic Offline Crowd Dynamics. From these events, it transpired that the common sentiment against the regime did not grow overnight but has been nurtured for years. The tipping point occurred when the people saw what others were thinking on social media, they realised that they were not the only ones and instantly became bolder.

The Internet Facilitates Repression Too. Offline surveillance might be difficult and expensive but there are countries that go to great lengths in order to enact it such as [61]. In countries such as Russia, Egypt and Tunisia, IP addresses were collected, active users were tracked and harassed.

Pressure Causes Adaptation. After enacting these censorship mechanisms, activists seek how to circumvent such controls in order to thwart surveillance.

Geography Matters. Since these regimes tend to block all sort of communication, the use of telecommunication infrastructure from neighbouring countries is vital. This is happening in Syria where the Turkish and Jordanian telecommunication systems are being used to convey news out of the country.

Think Small. One does not need massive infrastructure to communicate. When the Egyptian government switched off the cell-phone network, a device small enough to carry in a backpack was being used to reconnect the phones.

The New Threat Is Goldilocks Dictatorship. Smart governments learnt that the best suppression is to set up state-controlled intranets, which give the illusion of access to the outside world. This is what happened in Cuba and in Iran whereby the people only have access to a government-approved reality.

Beware Animal Farm. In the famous George Orwell book [215], the animals take control from the farmer and overthrow his dictatorship, only to realise that their fellow animal friends who take control can be worse than the authoritarian farmer. In essence, this is what we saw in Egypt [37] where the new government has been overthrown by a military coup.

Use the Internet to Keep Power the Right Way. Governments are now afraid of the Internet. In reality they should be if they misuse the internet. Rather than having an army of censors, they should employ a cadre of government officials reading posts and answering the legitimate grievances of the people. After all, governments are there to serve people and not to restrain them.

On a more formal note, we can see that Governments too adapted themselves to the traditional Web 1.0 by publishing information online, communicating via e-mails, etc. With Web 2.0, its a totally different story. Even though the technology has been around for almost a decade (practically since [232]) and considering that Wikis, Blogs, Instant Messaging, VOIP and Social Media makes it possible for information to be shared more effectively and almost for free, governments are finding it hard to adapt and provide Web 2.0 services to their citizens. Citizens can be considered as both the government's customers and its owners. Technology can be used to amplify and aggregate voices that used to be weak and muffled. Whereas before, voters could only write a letter to a newspaper in the hope of being published, now they have the publishing power in their hands through the use of social media application.

All the elected representatives need to ensure that citizens which might have remained unheard should be included in the running of the state. This can be achieved through online discussions about particular topics. Governments should not shy away from openly discussing policies or laws as they are being formulated. Legislators should engage in the online discussions through responses and in the end, they can derive a fair and unbiased summary of the views being raised. This process is already being used in a number of countries. Iceland [76] decided to crowdsource its own constitution. Finland [99] allows its citizens to crowdsource its laws. The list of similar initiatives keeps on increasing by the day.

However, with the rise of the new generation of digital natives, we have to see how they will be shaping our world. Online citizens spend their entire day chatting, working and playing alongside people based in different places around the world. A work does not have fixed office hours. One can be conducting business with someone in different time zones. Messages are sent back and forth through various mobile devices irrespective of whether he is in his office or no. Because of this, young people are considering themselves as being global citizens [158]. Apart from having a physical life, these citizens have an online life, as complex as the physical one and most of the time, much more engaging and colourful. They are active through the various crowdsourcing initiatives available online. They respect other people's point of view and if they don't agree, they simply explain themselves. They are not afraid of using other people's work but in the process, they acknowledge the author and give credit for the work used. The information at their disposal is massive and they're developing new skills in order to sift through the online resources and seek what they're looking for. Inclusion of others in their daily lives is a must, they have to inform their peers about their lives with status updates and share with them their daily events. However this is not just about engaging in active listening, they have to research the credibility of the sources, cross-reference all sources and decide accordingly whether to trust them or no. To do so, they have to master the use of computers, mobile devices and all the other technologies which allow us to surf the internet. But the digital natives of today do not just consume but they also actively contribute to the internet. They create blogs, wikis or simply post comments on any topic under the sun. In the process, they network with experts in the field via Skype, FaceTime or email. Thus, living in an online world where information is limitless and accessible, they feel that physical borders act more as a hindrance to their day to day needs. The App store, Netflix and Spotify to name a few, all limit their digital products to specific geographical locations.

Because of this, these citizens have been trying to break those boundaries imposed by national legislation. Back in 1996, John Perry Barlow, one of the founders of the Electronic Frontier Foundation[4], published "A Declaration of the Independence of

---

[4] https://www.eff.org

Cyberspace" a paper on the governance (or lack of it) of the internet. In 2006, we saw the rise of WikiLeaks [125] later followed by the Snowden case[126], whereby normal citizens are demanding more transparency from their governments and exposing questionable initiatives. In those years, we also saw the rise of Anonymous [166], a group of people who use computers in order to promote political ideals. The group gained worldwide fame when they conduced a series of DDoS[5] attacks on government agencies (such as those in the US, Tunisia, etc) and on large organisations such as PayPal, Visa and Sony. People seem to be opting more for an online kind of life whereby most of their interactions occur through digital mediums. We're already seeing people thinking of Cyber Marriage where a couple can get married over the internet. The couple together with their invites from all over the world can go to an online chapel chat room where an ordained minister is online and speak their vows. When the minister pronounces them husband and wife, the couple's avatar can kiss as they would do in real life. Some Universities are even organising graduating ceremonies online [242]. The list of examples can keep on going.

The question we should ask though is whether there should be some sort of government regulating these citizens or whether anarchy should prevail on the internet. If the internet turns into a limitless state, what will happen when people will start venturing into the extreme; such as the Tokyo man who decided to marry Nene Anegasaki, a virtual character in the Nintendo DS game called "Love Plus" [177] ? On the other hand, we have to ensure that the Internet maintains its independence and openness. A decade after the declaration of Independence of Cyberspace, when Barlow was interviewed in [88] , his views where somewhat more conservative. In the end, the Internet and its evolution is something which we still have to understand. We have to learn how to use and harness its power for the common good. We have to learn to handle it without suffocating it and its core values. We have to teach our digital natives how to make the best out of it in the hope that they will create a better digital future for all of us.

8.2 Digital Legislation

Legislators have been trying hard to enforce some sort of law over the internet, however this proved to be much more difficult than they envisaged. First of all there's the issue of territorial jurisdiction[6]. Any state has the right to exercise territorial jurisdiction over its own territory and over the people that live in that territory. The problem with the internet is that territorial boundaries are rather fuzzy since sometimes it is quite hard to trace who is doing what on the internet and from which server. However, to limit this issue, Country Code Top Level Domain (ccTLD) names have been created in order to facilitate the creation of Cyber-territories. The World Summit on the Information Society 2003 declared that "each government shall have sovereignty over its respective ccTLD".

---

[5] A cyber attack aimed at making a specific machine or network resource unusable by sending too many requests to the server thus causing its collapse.

[6] The relationship between different states.

Apart from these, according to Article 22 of the European Convention on Cybercrime[7], if a person commits a crime in country A from a computer in country B, both countries have the right to intervene[8]. With regards to the publication of harmful content on the Internet, the issue becomes somewhat more complex since a website can be accessed from virtually anywhere in the world. In [114], the Paris Regional Court found that Yahoo! Inc (operating from a US server) violated French criminal law by offering Nazi memorabilia which could be accessed from France. This case was rather straightforward since Yahoo! Inc is a well known company but [273] claims that due to the nature of the internet, it is not easy to simply link a webpage to a particular country. Thus, courts normally resort to other means when deciding on jurisdiction such as; the language utilised, the content and the publicity. If they do refer to a specific country or are clearly targeting a particular country, then it implies that it was intended to be retrieved from that country thus giving it a good claim for jurisdiction.

However, we have to be really careful that this issue of jurisdiction does not collide with the principle of internet freedom. There have been various drives, in the past decade, towards protecting this freedom such as:

- The People's Communications Charter[9] in 1999.
- The APC[10] Internet Rights Charter in 2006.
- The 2007 proposal for an Internet bill of rights which emerged out of the the Internet Governance Forum[11] in Rio.
- The Council of Europe Code of Good Practice on Information, Participation and Transparency in Internet Governance[12] in 2010.
- The United Nations Human Rights Council affirmed[13] that the same rights that people have offline must also be protected online".
- The Declaration of Internet Freedom[14] in 2012 created by activists and academics.
- The Reddit community too is working on a draft digital bill of rights [295].
- A number of EU states, including Spain, France, Estonia and Latvia, have already made the right to internet access a human right and countries such as Malta [325] are planning to introduce further "Digital Rights" in their Constitution.

On the other hand, there have been various threats (particularly in the past few years) towards internet freedom, such as:

---

[7] http://conventions.coe.int

[8] Council of Europe, Committee of Ministers, Convention on Cybercrime, Explanatory Report of 8 November 2001, paragraph 233.

[9] http://www.pccharter.net

[10] Association for Progressive Communications. Source: http://www.apc.org

[11] http://www.intgovforum.org

[12] http://www.intgovcode.org

[13] Resolution L13 The Promotion, Protection and Enjoyment of Human Rights on the Internet adopted by consensus by the Human Rights Council on Thursday, July 6, 2012.

[14] http://www.internetdeclaration.org

- The Combating Online Infringement and Counterfeits Act (COICA) in 2010 would authorise the Attorney General to order the suspension of a domain name if it was found infringing copyright law.
- The Stop Online Piracy Act (SOPA) in 2011 wanted to give the US Justice Department, the power to close down websites accused of copyright infringements without even giving them the right of defence. Copyright owners could seek compensation from companies that did business with the accused website and anyone who infringed copyright was risking a jail term of up to five years.
- The Protect IP Act (PIPA) in 2011 was a bill similar to SOPA which required greater court intervention.
- The Cyber Intelligence Sharing and Protection Act (CISPA) in 2011 aimed at sharing Internet traffic information between the US government and technology companies without the need of a warrant. This bill gives the government permission to monitor all the Internet activity of any individual.
- The Anti Counterfeiting Trade Agreement (ACTA) which started in 2006, was a multinational agreement aimed at placing the onus of guarding copyright infringements on the Internet Service Providers (ISPs).

Luckily, none of these bills succeeded mainly due to the massive protests [207] [244] [210] which were organised around the globe. However, it seems that notwithstanding these apparent victories by the Internet freedom activists, governments around the world still kept on pushing their plans of controlling the Internet. The biggest case is without doubt the [126] case where Edward Snowden revealed the internal operations of the National Security Agency (NSA). In particular, PRISM, a secret electronic surveillance program operated by the United States. Its aim is to locate and store internet communications from various high profile companies including Microsoft, Google, Yahoo, Facebook and Apple amongst others. From the dust which arouse around this controversy, it transpired that other countries such as the UK knew about its existence and were also using the information obtained from PRISM. In fact, [234] [336] claim that the British Prime Minister even defended the use of this data. This upheaval was soon followed by lawsuits by the American Civil Liberties Union and Freedom Watch USA against government bodies responsible for PRISM and against those companies responsible with disclosing the personal data. Professors from Georgetown University Law Centre [89] and Stanford Law School [95] claim that even though the program itself might be legal (since it was backed by the 2007 Protect America Act) it is probably unconstitutional since it violates the first and fourth amendment of the American Constitution. But the revelations did not stop there and apparently, even high profile people might have been targeted like the German Chancellor Angela Merkel. The situation is still very fluid but as German Justice Minister Sabine Leutheusser-Schnarrenberger said in [187],

```
The more a society monitors, controls, and
observes its citizens, the less free it is!
```

8.3   Online Politics

Behind online legislators, one find online politicians. In fact, politics too migrated
to the online world. [231] claims that the Internet changed the political landscape.
First of all, the time taken for good or bad news to reach the electorate is extremely
small considering that newsagents are constantly posting on social media and this
information is subsequently reported several times. The Internet and the electronic
payment gateways provide mechanisms for the efficient collection of small contributions from the grass roots aimed at financing the political parties. In fact during
the 2012 US presidential election, [32] estimates that Obama collected over a billion
dollars for his presidential election. Finally, as we have seen happening during the
Arab Spring ([148] [304] [167] [150] [116]) social media empowers small groups
of like minded people to organise themselves and create networks which are always
in contact with the top echelons of the political campaign. But to do so, campaigners
need talented people, this is where the power of the social networks come into play.
In fact, various organisations [17] [65] are using social networking sites to scout
people and identify talented individuals.

The first use of the Internet for political campaigning can be traced to 2002 when
the Dutch had their General Elections. [33] asses the role of political websites in
these parliamentary elections and to do so, it conducted an online survey of around
18,000 visitors. From this survey, it transpired that political websites manage to
involve a few people who remained aloof but the real success lies in reaching the
younger generation; a group of people made up of citizens who are not normally
politically active but who are avid internet users. However the survey also shows
that the use of these websites does not lead to more active engagement since their
primary use is to find information about the party, its organisation and its electoral
campaign.

A year later, we find the Belgian elections which have been analysed in [143].
During these elections, we can see the introduction of non-partisan Party Profile
Websites which allow the user to answer online policy questions and the system
will calculate the distance between his beliefs and a party's agenda. This system
is extremely useful in multiparty systems whereby it is hard to identify a party of
allegiance. However, the downside of such a system is that its users tend to be highly
educated young males. The system does not influence the voter in anyway and the
views expressed by the users are just a reflection of their beliefs.

The 2004 US presidential campaign experienced the rise of blogs. They continued their rise in popularity during the 2005 British general elections when they were
examined by [300]. A content analysis of over 300 blogs and 1300 messages was
performed. Through this work, it was concluded that blogs were still in their infancy,
even though there were many blogs, there were not many political ones and fewer
people accessed them on a regular basis. Most of the political blogs were biased
towards particular parties and offered observational analysis of what was happening
during the campaign rather than a discussion on the political factors effecting the
election. In essence, the blogs had no real effect on the campaign as they did in the
US presidential election held the previous year.

[282] ran a longitudinal study on the German elections of 2002 and 2005. The websites of the major political parties were examined and compared. It transpired that the main political party websites evolved over time, containing more information, interaction and sophistication. Smaller parties exhibited little or no change. Apart from this, it was noted that the content utilised for these websites was very similar to the traditional content used in other mediums thus there was no real adaptation for the web.

2006 saw the Italian Parliamentary elections which were analysed by [312]. His work was based on 27 websites which were analysed for information, participation and professionalism. The concluding results showed that most Italian parties were proficient in the provision of information to their constituents however they did not manage to maximise the potential of the Internet with regards to participation and mobilisation. The following year [313] analysed the French presidential and legislative election. It became evident that throughout that year, the Internet was gaining importance. The websites of the 12 presidential candidates and 10 national parties that participated in the different elections still lacked proper participatory tools. The online presence of parties and candidates exhibited significant differences. Also there was an evident gap between small parties and larger parties.

The main turning point occurred during the 2008 presidential election when according to [252] [323], social media became extremely important to gain an electoral advantage. The internet was not simply a media channel aimed at information consumption. Information systems where created to influence political scenarios since technologies were developed which allowed users to interact and collaborate with each other. These technologies changed the nature of competition in politics and complemented traditional media. People expect politicians to be connected to them in some way or another. They want to feel them close because this link tends to empower them. Politicians are the ones who eventually make the decisions (if they are elected) thus a closeness to a politician is interpreted as a way of having a say in that decision being taken. Social media allows this to happen in real time since these politicians are only a message away. Closeness also implies a relationship of some sort so if the politician regularly consults his constituents, he is being seen as accessible and will be remembered as being in touch with them. When they post messages on their wall or update their blogs, they are being seen as consulting with their constituents through the comments which are posted as a reaction to the particular post. Such an approach means that the constituents now have a voice. Through this voice, they feel as if they are actively contributing to politics and get the impression that what they wrote is being given the attention due. Finally, the interaction between the politician and his constituents will result in a show of empathy where the politician is being sensitive to the needs of his constituents whilst also explaining better the political decisions being taken by the party he represents.

Notwithstanding the fact that technology grossly intensified the democratic process, it did not change its fundamentals. If we have a look at the 2012 presidential election in the United States [32] [331]we find that President Obama (who was seeking reelection) decided to opt-out from the public funding offered by the state to run his campaign. Instead, he went for crowd-funding[15] and collected a small amount from many donors. Even though we experienced a massive online campaign, President Obama's team still had to organise face-to-face meetings, conventions and other events where people had direct contact with their candidate.

According to [314], the electronic campaigning tends to reinforce political attitudes but not change them. [168] and [169] reached the same conclusion when analysing the 2007 Australian federal election. Although the internet was extensively used by Australian campaign managers before the elections, when considering that YouTube videos and Facebook pages were being reported on local news, the analysis still shows that the major gain in political support cannot be attributed to Web 2.0 but to a change in leadership which occurred months before. [197] goes a step further by claiming that the Internet was only used like a mass media communication channel rather than to engage in listening, consultation and dialogue. More worrying was the fact that he got the impression that these channels were being controlled by gate keepers and image makers who were more interested in the public relationship of their candidate or party than in actual politics. The messages being sent by politicians to their constituents needs to be received and accepted if they are to be effective. This is why a lot of work is going on about finding new approaches to improve receptiveness and acceptance. When this barrier is overcome, diffusion automatically increases as citizens start reposting the message touted by their political party.

An interesting phenomena which evolved in the past years is the creation of alternative parties. Parties which were created in response to various crisis brought about by conventional parties. In 2006, Rick Falkvinge created the Pirate party in Sweden. This phenomena soon spread world wide and today we find more than sixty Pirate parties around the globe. Some were very successful by electing candidates in local and national elections whilst in 2009, the Swedish party, even managed to elect two members in the European Parliament obtaining 7.13% of total votes. The original idea behind the party was to create a movement aimed at reforming the intellectual property rights, copyright laws, strengthening civil rights, create a transparent government, guarantee speedy fair trials, fight for freedom of speech and guarantee anonymity in communication. It is interesting to note that the party operates through internet communication and interaction between its members. In fact, the party's primary communication tools are Google, IRC, Skype, Twitter and Facebook. It can be defined as being the first Cyber party in history, a sharp contrast from the traditional political parties we're used to.

8.4 Cyber Crime

Even the underworld of crime migrated to the Internet; these days we use the term Cyber crime to refer to a crime conducted via the Internet or through a computer network. It can be defined as being an offence targeted at individuals or a group of

---

[15] Crowdfunding is the collective effort of individuals who donate money in order to support initiatives usually via the Internet.

people in order to cause some form of damage [159]. Because of this, law enforcers need to utilise new methods when targeting cyber criminals (such as [5] which uses Social Media to prevent crimes). Cyber crimes can be categorised into three main areas; individual, property and government crimes.

Individual crimes effect people and their personal life. This includes cyber stalking, pedophilia, trafficking, etc. These crimes seem to be on the increase, in fact they affect around 1.5 million people a day [286] costing annually around \$100 billion. A figure which is going to double in the coming years especially with the proliferation of personal devices such as tablets and smart phones. Various law enforcement agencies have joined forces in order to prevent these crimes such as the European Cybercrime Centre (EC3) at Europol. The age of the victims is drastically going down considering that most young children in the developed world have access to a computer or to a digital device capable of connecting to the Internet. [106] states that the majority of offences involving children does not stem from adults preying on the children but from their own peers. Half of the kids at kindergarten level interact with other people online in some way or another and half of these kids declared that they have been exposed to some content which makes them feel uncomfortable. What's worrying is that at such a tender age, only half of the parents supervise their kids while they are online. As kids grow older, surveys reveal that parental supervision tends to decrease. During the same period, cases of cyber-bullying start to appear. Further still, these children start meeting strangers online and one-in-ten of those kids have also been exposed to private things about someone elses body. These are also the years when children start posting things online with complete ignorance of privacy issues. They are not aware of the extent of their actions and do not realise that their information is being splashed on a world wide billboard. During this phase, children start experimenting with software piracy as well. Even though websites such as Facebook have age restrictions in place, the truth is that more than 40% of the children under the age of 13 own an active Facebook account. By the time they reach secondary school, most of the kids report almost no supervision from their parents. The issues we mentioned earlier keep on intensifying and some of the children even agree to meet strangers in person. This can obviously be very dangerous and we have heard quite a number of cases of Cyber killers [324]. The issue of Cyber bullying [41] [191] is also on the increase and can reach almost 20% of adolescents. A person is bullied when he is repeatedly exposed to negative actions on the part of one or more other persons and he has difficulty defending himself. On the internet, the issue intensifies itself because it is extremely easy to create fake profiles thus a person might get the impression that he is being targeted by a large group of individuals. In recent years, we've witnessed quite a number of incidents concerning adolescents who eventually committed a suicide and most of them stemmed from cyber bullies who used a website called Ask.Fm [28]. In fact, after these episodes, the families of these victims started a crusade to shut down this site.

Notwithstanding these incidents, we should not discount the dangers associated with adult predators. It can be seen in [275] [334] [97] [112] that there exists groups of organised pedophiles aimed at sexually grooming and preying children. It is quite interesting the case of Wonderland, a pedophile secret playground hidden in Second

Life (SL)[16] whereby avatars could have sexual encounters with children in the playground. To counteract these organised groups, we have the parents, whose monitoring techniques are not enough to help keep their kids safe. There is simply too much content being created by the the children and there are too many predators lurking around the net for parents to keep track of what's going on without help from the authorities. We expect this situation to only intensify in the coming years as more social networks develop and more kids get involved.

Even though most of the crimes mentioned so far targeted children, we have to keep in mind that even adults are at risk. Apart from individual crimes (as in the case of children) they can also risk being victims to property crimes. Cyber world criminals can also resort to stealing and robbing. In this case, they can steal a persons bank details and use all of his money. They can run a scam to get naive people to part from their hard earned money by using social engineering techniques. Malicious software can be used to gain access to an organisations website or disrupt the systems of the organisation. The malicious software can also damage software and hardware, just like vandals damage property in the offline world.

A growing trend on the Internet is the setting up of online markets selling virtual objects [185]. People buy houses on SL, cows on FarmVille, virtual Nike shoes for their avatars and all sorts of virtual objects. These objects (similar to real life) have a virtual representation of the object itself, can be bought using a buying gesture and can be exhibited (by showing them around in case of a house, using its service in the case of a cow or dressing up the avatar in the case of the shoes) so that other people can see them. This is an online marketplace where real money is being exchanged thus making the virtual objects valuable, desirable and as a consequence can lead to someone else committing a crime to obtain them. In 2005, the China Daily newspaper reported the case of Qiu Chengwei, an avid gamer of the popular online game Legend of Mir 3, who stabbed his friend Zhu Caoyuan repeatedly in the chest after he was told that Zhu had sold a virtual weapon which he had lent him. Qui went to the police to report the theft but was told that the weapon was not a real property protected by law. The lack of legislation to protect online gamers unfortunately instigated Qiu Chengwei to resolve the matter using a real knife.

Interaction in virtual worlds and chatrooms can also be the reason behind passion crimes. As a result, people face consequences for their actions in their real lives [94]. For example, a woman was arrested in Japan after killing her avatar husband in the virtual environment Maple Story, an online MMORPG (Massively Multiplayer Online Role Playing Game). She is now facing a 5 year prison term together with a $5,000 fine. Another woman in Delaware, US was charged with plotting the real-life abduction of her ex-virtual boyfriend which she met through SL. Because of these incidents, different countries are now enacting cyber crime legislations such as [57] in order to protect their citizens.

Even though it is not as common as the previous two categories, crimes against governments are on the increase [72]. These are normally called cyber terrorism and they're aimed at wreaking havoc or cause panic amongst the citizens of a state.

---

[16] An online virtual world developed by Linden Lab accessible at
   http://secondlife.com

[330] [160] claim that all the government institutions, from Military installations to water systems can be targeted hence national security must guarantee their protection. Further still, Richard Clarke the former chairman of the White House Critical Infrastructure Protection Board claims that whereas before, one could estimate the armaments withheld by a hostile state or a terrorist organisation, in the case of cyber warfare, the enemy's strength can only be estimated after he uses it.

There are normally three types of cyber terrorist attacks [259] which include:

Physical   attacks which make use of conventional weapons (such as bombs, firearms, etc) to destroy physical hardware.

Syntactic   attacks make use of computer viruses which damage of destroy a computer network. The software used might include worms, trojans and DDoS attacks.

Semantic   attacks which cause a computer system to produce unexpected errors and results thus mining the confidence of its users.

Cyber terrorism is becoming much more attractive to terrorists than traditional methods [52]. It is more cost effective since it does not require expensive armaments and the cost of computers, phones or broadband Internet connections is much cheaper. The basic devices needed to perform an attack can fit in small spaces and it passes through security checks. A cyber terrorist might strike from anywhere in the world, from the office next door to a tent in the middle of the Sahara desert. Thus the physical and psychological training aimed at helping the perpetrator evade capture is eliminated since everything is happening without his physical presence. He need not work on his own but might be operating with various other cells, dispersed around the globe. This setup is rather dangerous because it is hard to trace, considering that when active, these cells make thousands of simultaneous attacks using infected 'zombie' computers worldwide as happened in the case of the Sony attack [184] amongst others. The information needed to operate such devices and plan an attack is freely available online. The attacker can enjoy virtual anonymity since it is quite hard to be tracked and at the end of the attack, he does not sacrifice his life as in the case of a suicide bomber. The attack is also much more effective because while a suicide bomber may destroy a building targeting tens or hundreds of individuals inside, a cyberterrorist attack has the potential to kill or injure millions of people. Just think what would happen if terrorists manage to control a nuclear power plant! Further still, due to our increasing dependency on information technology, a cyber attack might put an entire nation on its knees.

In essence, we can say that every political and military conflict these days always includes an element of cyber warfare. [190] further states that in this day and age, a victory in cyber war almost certainly reflects a victory on the ground.

Chapter 9
Future Trends

Predicting the future is something we all do as part of our daily life. In fact, we do not really live in the present, because as soon as you think about it, it is already in the past. Thus, most of our thoughts abound around future events. However, correct predictions are not easy. First of all we need past experiences through which we can extract a model of how the world works. Once we have that in place, we need to make assumptions and then infer new knowledge from the combination of the world model and those assumptions. This is what we will try to do in order to predict what will impact our life in the future. Having said that, the major technological advances which we have experienced in the past decades were hard to predict and in some cases, the ideas behind the technologies we have today were secluded in science fiction novels. Just think about computers, the Internet, social network, etc. The fact that technology is reaching new boundaries is automatically unleashing new possibilities for everyone, not just for programmers. Further-still, there is a drive[1] for more people to become programmers since in the future, our dependency on technology will most certainly increase and as a consequence, we need people capable of programming that technology. Also, people have been predicting [209] the rise of Artificial Intelligence. With multi-core chips being planted in mobile devices, the setting up of clouths[2] and quantum computing on the horizon [266], the era of strong AI[3] is about to begin and the world we know today is going to change forever.

9.1 The Evolving Internet

The internet is without doubt one of the biggest innovations of the past century. It easily surpasses other inventions for the simple reason that it is the first time in human history that humans managed to create a library of global scale where all the

---

[1] http://code.org
[2] Coulds of Things.
[3] AI that matches or exceeds human intelligence.

© Springer-Verlag Berlin Heidelberg 2015
A. Dingli and D. Seychell, *The New Digital Natives*,
DOI: 10.1007/978-3-662-46590-5_9

past, present and future wisdom is being recorded and can be easily accessed by anyone. Having said that, not everyone has access to the Internet and in fact, only 2.7 of the world's 7.7 billion population had Internet access in 2013. Thus, one of the ginormous tasks which Internet Service Providers are facing is how to provide Internet to the entire globe. At the moment, there are two competing initiatives, one of them is Internet.org and the other if Google's project Loon. Internet.org was spearheaded by Facebook and includes big names such as Ericsson, Nokia and Samsung. The idea behind it is to make mobile access; less expensive, designing more affordable mobile devices, developing simpler apps and create Internet technologies to reduce the amount of data that must be transmitted. In so doing, the costs associated with accessing the internet are reduced. Loon on the other hand is all about creating networks over large areas, using high-altitude (around 20 Km) balloons placed in the stratosphere, which promise to deliver 3G connectivity to remote areas.

Apart from connecting the world, we need to upgrade its infrastructure. We've heard in the past of terms such as the Information Superhighway, Internet 2, etc. Now we're looking at extremely fast speeds, not only in physical but also in wireless links [69]. This is necessary because people are already consuming 50 Exabytes of data per month according to [154] and this is estimated to double in a few years time. In the case of mobile devices, global mobile data traffic will increase 18 times from 2011 to 2016. Broadband will be accessible to everyone who owns a device being it a networked machine or a mobile device. This will pave the way for a new generation of multimedia, one which is much more rich and powerful and which allows for new ways of interaction. Apart from this, various devices can connect to different things. By things, we mean anything which can be connected to the Internet being a TV, a microwave but also a simple lightbulb. In the future, anything can be connected to the net and in fact, it is being referred to as the Internet of Everything. But even though the internet is expanding into consumer items, most enterprises and technology vendors are not ready for such a change, thus it is expected that it will take some time until we start tasting the benefits of it.

Intelligent systems will work on huge datasets known as big data. This immense stream of information is harvested from both internal and external sources. Organisations today have the luxury to analyse that data, transform it, use it for decision-making, discover new insights, optimise their business and innovate in such a way which was unthinkable a few years back. Big data creates the real information economy whereby information is now being turned into revenue which will eventually accelerate the growth of these businesses and that of the global economy. These huge datasets allow companies to look ahead using predictive algorithms and the companies who manage to master this technology will have a competitive edge over their peers. The demand for big data is so large that it is estimated to generate more than 4.4 million jobs globally in IT over the coming years. For every person working in big data, three additional jobs outside of IT will be created thus bringing the figure up to almost 18 million. Ironically, only one-third of these jobs will be filled according to Gartner because there is not enough talent in the industry. Organisations will need people skilled in traditional fields such as data management, analytics and business expertise but also in non-traditional skills such as artists and designers for data visualisation.

To run these algorithms over these massive datasets, traditional computing is not enough and a lot of organisations are turning towards cloud computing. Large companies such as Amazon and Google are re-inventing the delivery of their IT services in terms of speed and agility. But cloud computing is not something restricted to the enterprise and in the future, we can foresee the era of personal clouds. The device used to connect to the cloud will become irrelevant because in a personal cloud, all the information, apps, etc pertaining to a person are stored somewhere online. The device simply becomes a window through which one can access his own personal profile thus making it possible for a user to access this data through multiple devices. This demand for on-the-fly data will strain our global infrastructure and the need to manage bandwidth will automatically increase. Managing clouds will become more complex, users would have to be incentivised in order to minimise the load and intelligent machines will become vital towards ensuring a fair distribution amongst all users whilst guaranteeing the quality of their experience. This will lead to a global restructuring of the IT services market, cannibalising smaller players in the process. The strain on server side technologies is also inevitable since most of the processing will occur over there. The solution to this problem will probably result in hybrid clouds where personal and external private clouds are used, thus sharing the load between the organisation's own servers and other external ones. This will be attained through service brokerages where the system will decide what can be computed internally and what can be farmed out to other organisations.

9.2 Intelligent Systems

To effectively deal with the world around us, we need to create intelligent systems. First of all, the technological advances of these past years are making it possible for researchers to push the boundaries of Artificial Intelligence further. Apart from that, the proliferation of personal and wearable devices increased the consumption and creation of digital information. According to [111], the digital universe in 2005 consisted of 130 exabytes. By 2020, it will reach 40,000 exabytes and after that year, it will double every two years. The figures are rather impressive and show that the digital universe is growing so much that it is virtually impossible for a human to handle all this influx of data. Thus we need intelligent systems to help us handle it.

The front runners in this quest seem to be top firms which are harvesting big data, such as Facebook. In fact, in 2014, Facebook announced the creation of a new Artificial Intelligence lab aimed at designing deep learning systems. This initiative is extremely valuable for the social network because it will help to automatically identify people in photographs, tag them using the correct name and instantly share the photos with friends or family. Such a system will also profile users by taking into consideration personal relationships, daily activities, voting preferences, purchases and all sorts of online information. It will then create a mathematical model which would allow Facebook to predict which posts are the most interesting for the user. However, apart from the personalisation and adaptivity of these systems, another important aspect is the user interface. Users have to feel in control and empowered to perform whichever task they require. The recent drive towards touch interfaces has made a huge difference. Essentially, people only need a finger to operate a system on one of these devices. This automatically lowered the entry barrier thus making it possible for young toddlers [271] to start using these devices from a tender age. Whilst touch is perfectly fine for simple interaction, it might get difficult to use when conveying complex thoughts. Because of this, we need other inputs and the most natural, for most humans, is without doubt natural language. The rise of cloud computing has made it possible for voice enabled systems to analyse complex vocal messages. In particular, we've seen the introduction of SIRI[12][4] in the mass market whereby users can issue voice commands to a mobile device, the system understands the sentence and answers back using either an action or a voice message. Google too entered the race with Google Now[5], an advanced digital personal assistant capable of sifting through the daily routine of the user and provide him only with the details required before they are actually requested by the user. All of this can be controlled using Google Now's voice commands which has similar capabilities to SIRI. Google is also integrating its voice search in other products such as Google Maps and YouTube. Microsoft too entered the race with /Cortana (a system similar to Siri) and the Microsoft Skype Translator. The Skype add on is capable of simultaneously translating between one language and another during a live conversation thus breaking down language barriers between people. This technology will be deployed towards the end of 2014. The examples mentioned earlier clearly show that voice is going to play a very important role and we expect more devices to adopt sophisticated voice interfaces in the near future.

Notwithstanding this, Google Now introduced a new paradigm whereby the ultimate goal of such a system is to have information intuitively delivered to the user's devices (thus shifting away from having to actively request data) in a passive way before it is actually requested. This promotes the concept of instant information whereby the system learns to distinguish which information is required by the user and simply presents it to him. To do this, we need more sophisticated AI on our desktops. In a study conducted by [329], it was found that the average UK worker sends and receives 40 emails a day whilst 1 in 12 people receive more than 100 emails daily. In a separate study, [60] found that email is the second-most time-consuming activity for workers taking around 28% of their time. This function can be taken over by intelligent systems whereby smart software will be capable of drafting appropriate responses (scrapped from the Web or from other databases containing relevant information). AI software can be used to evaluate the various alerts sent from chats, phone calls, etc and filter them out whilst also deciding whether to interrupt the user with the alert or suppress it. In essence, intelligent systems can be used for the most mundane tasks thus providing users with huge timesavers which can be allocated for other things.

---

[4] SIRI is a system developed by Apple which allows users to use their own voice to send messages, schedule meetings, place phone calls and much more.

[5] http://www.google.com/landing/now/

9.3 Machines

Even though software is becoming more powerful, its power is ultimately dependent upon the underlying hardware. In the past years, we have seen a closer symbiosis between the two. The days of rigid machines is rapidly changing and hardware is becoming much more flexible. It is enough to have a look at Google's Project Ara[6], a project aimed at creating an open hardware platform whose ultimate goal is to create modular smartphones. The idea is very similar to lego blocks whereby any component in a mobile phone can be bought separately and installed. To upgrade the phone processor (something which is currently impossible with existent phones) is simply a matter of unplugging the current processor and inserting a new one. To change the battery, screen, sensors and all the other components is essentially the same process. Because of this, it is not surprising that Gartner[7] estimates that spending resulting from the proliferation of smart technology is to increase by 25% in 2014. The increase is not only restricted to smart phones. Most new devices (ranging from vending machines to medical devices) have some sort of software embedded in them which links their sensors to the internet thus creating a constant stream of data. This will lead to a significant rise in costs when one considers that machine-to-machine communication is expected to increase rapidly in the coming years. The change will not only happen at corporate level (as in the current case of big data) but at all levels. Simply buying a drink from a vending machine will involve having your smartphone converse with the vending machine, passing over electronic money and as a result, ask the machine to dispense the selected product. Essentially the possibilities are endless.

The coming years will also see a major shift towards mobile personal devices. A person is expected to own several of these devices probably in different shapes and sizes, and to carry them with him based upon their usage. One of these devices is without doubt electronic paper [287][216] which has all the properties of traditional paper with the added benefit that its content can change electronically. Thus, the era of digital newspapers is slowly beginning.

Another device which will probably gain importance is the Virtual Reality (VR) headset. Even though this technology has been with us for the past decades, the recent acquisition by Facebook of the Oculus VR Headset [121] fuelled new speculations. Facebook believes that some of the functions found in social networks can be ported to a virtual reality world.

Apart from these technologies, there are other devices and by 2016, it is estimated that more than 1.6 billion smart devices will be purchased globally. These devices are also making significant inroads in the workplace whereby 20% of sales organisations are already using tables as the primary tool for their field sales force and in the coming years, almost half of the workforce will be mobile. By 2018, the use of tablet-like devices in the workplace would have risen to more than 70% of the workforce.

---

[6] http://www.projectara.com
[7] http://www.gartner.com/

We've already seen in the past years a shift from the almost defunct Blackberry towards tablets. This means that enterprise software has to be updated and become mobile since the nature of the organisations is also rapidly changing. People are getting used to a society whereby information is at their fingertips when and where they need it. Because of this, they will be less inclined to use a service which forces them to go to a specific physical location at a specific time (like most offices today).

The challenges of software designers will be various. First of all they have to provide applications which deliver personalised and context sensitive services. Secondly, they need to manage the diverse mobile devices which are available. Luckily for developers, the online scene (on the software front) seems to be converging towards two main technologies; JavaScript and HTML5 while the browser is touted ·to become the mainstream enterprise development environment. This will probably shift the focus from the software per se towards the user experience whereby voice and video will keep on gaining importance. Internet applications need to leverage on the intent behind every action and in so doing, push the changes which the user needs in a transparent way. Desktop applications will therefore continue to shrink whilst the importance of mobile apps will keep on increasing. These devices will also bring forth a massive increase in electrical usage as a consequence. In order to avoid having a jungle of power cords adorning the home, devices will make use of wireless power [38][291]. This will charge the devices by simply placing them on or near to a charging surface without using a physical chord.

Apart from the personal machines which we mentioned, there are other smart machines which will take over our lives and they are expected to be the most disruptive innovations in the history of IT. Transportation machines are becoming vital in some cases. We've already seen drones being used in modern warfare [45], journalism [120] [319] and for all sorts of applications. So far, their usage has been restricted to tasks which were hard or risky for human beings. Now they are infiltrating mundane tasks as well such as delivering parcels [127] or running entire warehouses [194]. Machines will also take us around and in this regard, there are various researchers working in order to create autonomous vehicles[204]. Google seems to be at the forefront of such a technology and its cars have now passed the milestone of more than 700,000 autonomous miles.

In the field of consultancy, machines such as the IBM Watson [119] [7] (a cognitive system capable of processing information in a similar way to a human using Natural Language Processing, hypothesis generation, learning, etc) will become the smart advisor of the future. Other similar super computers coupled with cloud technologies will be created. However the real breakthrough will occur when we manage to harness the power of Quantum computing. This new type of computing is based upon the laws of quantum mechanics. Qudits [222][8] will replace the traditional bits used in today's systems. Since these bits can be in different states at the same time (in contrast to the 1 and 0 state of traditional computing), this will result in massive speed improvement. Further still, a recent innovation called Quantum spintronics [15] uses the spin state of electrons to transmit information in two directions at

---

[8] A multiple state quantum bit.

once, further reducing the size of electronic circuits required. In 2014, Google announced that it is using a D-Wave [39] [180] machine which is the first commercial quantum computer available. In the near future, we will probably see other quantum systems on sale.

9.4 Entertainment

The entertainment sector is also set to pass through a revolution in the coming years. If we look at TV, we are currently seeing various innovations such as Smart TV which essentially connects a normal TV to the internet. Some of them also give users the possibility to install applications such as games or even popular services like Netflix, Hulu and Amazon Instant Video. However the biggest issue is how to connect the two together at a deeper level. Various approaches used today are rather half baked whereby online information is displayed alongside normal TV channels. In this case, the level of integration is still rather superficial and there is much more which can be achieved. Most of the problems arise from the fact that the data being transmitted through the normal TV channel is not enriched with semantic information thus it is hard for smart algorithms to interpret the content and match additional information to it. Notwithstanding these issues, the shipment of Smart TVs is expected to explode in 2014 reaching around 123 million units. Competitors such as Samsung, Panasonic and LG are fighting towards achieving supremacy. As a result, the past years have seen the popularisation of High-Definition Television (HDTV), 3D TVs and now 4K TVs[9] with the prices plummeting. However the TV did not reach the top notch in its evolutionary scale and in the coming decade, we expect to see the introduction of Holographic TV [284][283] and other advancements.

Apart from buying Smart TVs, one can simply install a set-top boxes such as Apple TV[10] and Google TV[11] to his existent set. This tends to offer similar features to those offered by a Smart TV however there is no lock down since one can easily retain his existent TV and simply change the set-top box. These set-top boxes offer various features but the maturity level of the technology is still low. In the near future, we expect to see:

- Better user interfaces which utilises other devices such as smart phones or other mediums such as voice (rather than the archaic remote controller);
- Improved software developers kit (SDK) which allows programmers to create their own apps. This has to provide new features which harnesses the massive displays available and which exploit the different forms of inputs available on a TV.
- Fresh content capable of targeting a diverse audience but apart from purchasing it, new content has to be generated as well.

---

[9] 4K TV or Ultra-High-Definition Television deliver four times the picture resolution of HDTVs.
[10] https://www.apple.com/mt/appletv/
[11] http://www.google.com/tv/

- Increased machine-to-machine communication with regards to home automation. In fact TVs are being earmarked [46] as being the hub of home automation whereby all the various devices connected to the home network can be controlled through the TV.
- New domains where set-top boxes can be utilised such as in cars (for back seat entertainment) or in any other commercial vehicles (busses, trains, aeroplanes, etc).

Apart from home devices, another growing field is wearable technologies. The most obvious which have already reached the market are smart watches; which are watches capable of displaying email, text and social media updates at a glance. Samsung recently unveiled its Galaxy Gear Watch[12], Apple just presented its Watch[13] and Google launched its Android Wear[14]. Considering that smart watches have a very limited display, the problem arises when one needs to display information or request it. To solve this issue, apart from having a close integration with the smart phone, Google's watch is expected to come with Google Now, a technology which seamlessly provides relevant information before the user actually requests it. With regards to input on such a small display, the most obvious technology seems to be voice control. However to make this possible, voice commands need to be much more accurate than they are today. Notwithstanding this, smart watches should not be just a replacement for smart phones and they need to provide additional services to be effective such as heart rate monitoring, etc. Such sensors have already been integrated in fitness wrist bands[15] but they still need better integration with smart phones.

Another important wearable technology likely to become popular in the coming years is eyewear technology such as Google Glass[16]. Glass allows users to view two main pieces of information, one which is based upon the context and the other which is context independent. Imagine a user looking through Google Glass towards a shop front, the technology can give the user a glimpse of the various discounts offered by that particular shop. Thus in this case, it is providing contextualised information. On the other hand, a user on a bus might be reading the news or even email through his glasses. In this case, the information provided is not bound to the context of where the user is located. All the information shown to the user makes use of Augmented Reality technology whereby the virtual objects are superimposed upon real objects (through the glass). Commands are sent to the device either through gestures or voice commands. At the moment, the technology is only available to selected users and we expect a quick adoption in the coming years once the technology is commercialised. An advancement over this technology is Digital Contact Lenses[294] whereby the glasses are replaced by contact lenses. However the technology is still

---

[12] http://www.samsung.com/uk/consumer/
   mobile-devices/galaxy-gear/
[13] https://www.apple.com/watch/
[14] http://www.android.com/wear/
[15] Such as http://www.fitbit.com
[16] http://www.google.com/glass

in its infancy and will take some time before it matures. Another biometric sensor which has been successfully used in a digital device is the fingerprint scanner which has been integrated in the iPhone 5S. Although it is mainly used to unlock the device, we expect more applications based on this technology such as access to other systems, home security and even to validate purchases.

The most interactive form of entertainment is without doubt gaming. After years of expectation, we've seen the launch of the next generation of consoles mainly the Xbox One and the PS4. However the innovations on these two platforms were not massive. As expected, the consoles kept on increasing their social media integration thus getting players to interact whilst playing together. Players can easily compete together, work in teams in order to achieve massive goals (as can be seen in [54] and [279]) and if they're successful, they can also have people following them. Gaming is also becoming a massive business so much so that in 2014, the revenue from games and game content has reached the $ 24 billion mark thus surpassing the movie industry. The IGN entertainment website recently reported[17] that the prize pool for the international game championship of the online game Dota 2 started from $ 6 million and has now reached[18] the $ 11 million mark. Controllers are also becoming extremely sophisticated. In particular, the trend seems to be towards contact-less tracking of movement and gestures thus making the gaming experience more intuitive.

Not withstanding this ginormous growth, the biggest growth we should be seeing in the coming years is in the field of gamification[19]. Gartner estimates that 40% of the top 1000 organisations in the world will make use of gamfication to transform their business ventures. This shift is happening because organisations need to optimise their workforce. They intend to do so by introducing gaming elements such as engagement, introducing direct visible links between actions and business outcomes, creating a feedback loop for their workers, measuring and incentivising performance, etc. Essentially they will apply the same techniques which game designers use in order to keep their employees interested and help them achieve their corporate goals. The gamification marker is expected to reach the $2.8 billion mark in the coming years.

9.5 Commerce

In the previous section, we've seen the rise of wearable technologies. However we've looked at them from the functional viewpoint. In this section, we'll be looking at wearable technologies from a different angle, more as a fashion statement. In fact, smart electronics are expected to make their way into shoes, tattoos and all sort of apparels. This industry is estimated to reach the $ 10 billion mark very soon.

---

[17] http://www.ign.com/articles/2014/05/20/
dota-2s-international-prize-pool-exceeds-6-million
[18] http://www.dota2.com/international/compendium/
[19] Gamification is the application of gaming elements (such as points, leaderboards, etc) to areas which are not traditionally related to game (such as work or education).

This category does not only involve physical apparels but also virtual ones. A rather important marketplace for virtual apparels is Second Life. It is a known fact that large companies such as Nike regularly advertise virtual products (on this marketplace) as replicas of their physical shoes and sell them at the same price of a real shoe. Second Life users then use the virtual shoe to adorn their avatar. It comes to no surprise that the most popular wearable smart electronics in the coming years will be released in the field of athletics (shoes and fitness tracking), communication devices (such as headsets or earpieces) and insulin delivery for diabetics. But these devices are not only used as fashion statement and they do provide their users with real benefits. Fitness bands such as Nike+ FuelBand provides the user with data analysis used to extract important insights about the health of the user.

Another important technology which will revolutionise online commerce is without doubt 3D printing[20]. The past months have see a dramatic fall in pricing and we expect to see the appearance of the first cheap commercial 3D printers soon. This sort of printing will bring forth various benefits. Places such as China will eventually lose their competitive edge due to high costs of shipping and overseas contract management. Thus, 3D printing will slowly take over reaching a 200 % increase in the coming year. This is due to the fact that 3D printing has became a real, viable and cost-effective means of producing cheap prototypes and short-run manufacturing. The applications of such a technology are various.

- [193] proposes the printing of nutritionally balanced food.
- Rather than having a 4D video of a foetus, parents can now request a 3D printed model of their unborn child.
- [221] proposes the printing of organs, who knows maybe even bionic ears [202].
- Honda is using this technology to print prototypes of their concept cars.
- The activist Ral Aguayo-Krauthausen used 3D printers to build portable ramps for his wheelchair to help him go on pavements.
- People have used these printers to create 3D selfies which are used as a decoration around the house.
- Architects are using them to create 3D models of houses and some of them are also experimenting with the possibility of constructing a massive 3D printer capable of printing a large scale house.
- Children can use 3D printers to create their own customised toys such as Monstermatic[21].
- Forensic experts are using 3D printers to create replicas of crime scenes which are then used in court as part of the evidence.
- Customers can buy jewellery online, customise it (size, colours, etc) and simply print it, thus saving them having to go to the physical shop or from waiting for it to arrive by post.

---

[20] The process of laying down successive layers of thin material on top of each other in order to create a physical object from a three-dimensional digital model.

[21] https://www.kickstarter.com/projects/claytonmitchell/monstermatic-the-first-3d-printing-game

The possibilities are practically endless.

Another innovation on top of 3D printing is 4D printing. The two technologies are very similar however, 4D printing has the added capability of printing a 3D object which is self reconfigurable and reprogrammable (after it is printed). Thus, it is a non living object which can change its shape and behaviour over time without the need of any source of energy to do so. The only requirement so far is to add water to the object and it starts unfolding. Imagine printing a flatpack structure and by simply adding water, it will turn into a rocking chair. Imagine having smart beams which are used to automatically construct bridges. Essentially, it is very similar to how proteins work in our body. Whilst a lot of people are excited with the possibilities offered by 4D printing, the technology still has a long way to go but we expect to see more advances in the near future.

9.6 Human Element

Notwithstanding the various advancements in technology, they will all rotate around the most important component; the human element. Social networking is expected to make huge leaps in the coming years especially in the domain of consumer social networking. This concept will probably be at the top of most marketing strategies in the coming years. In fact it is estimated that in the coming three years, the 10 topmost organisations in the world will spend more than $1 billion on social media alone. This is due to a shift we've been noticing in the past years whereby social media was on the periphery of most organisations but it is now moving towards the core of their business operations. Whilst industrialised IT solutions in the West will probably lead to reduce the IT staffing, consumer social networking and other innovations mentioned earlier are expected to eventually create a large number of IT jobs in the far East. This will create a disparity between the different economies of the world where job mobility and outsourcing will most likely surge. In fact, it is expected that the European Union will issue some directives in order to drive legislation towards protecting jobs and reducing offshoring by at least 20% in 2016. This should create new opportunities for tech companies to invest further in lower-cost parts of Europe. However, there are still shortages which need to be addressed. Europe is expected to suffer a shortage of around 1 million IT experts in the coming years. That is why in recent years there has been various initiatives to teach children how to code. One of the most successful initiatives is Code.org[22] managing to make 38 million people worldwide conduct at least one hour of code.

The rise in mobile device usage (as seen in other sections) will probably lead to a new change in the workplace. Before, there was a strict separation between corporate devices and personal ones. A number of restrictions where imposed such as movement of devices between work and home. Restricted access to certain websites on work devices, etc. In order to reduce corporate costs, some enterprises are considering adopting a Bring Your Own Device (BYOD) initiative whereby employee-owned devices will be allowed at the workplace and used for work related purposes.

---

[22] http://code.org/

Obviously, this implies that clear policies should be established in order to decide what is acceptable and what isn't. Policies should be carful to protect the individual's privacy whilst also securing the workplace from corporate leakages of sensitive information or malware attacks (since employee devices are twice as vulnerable than corporate devices). These threats are extremely real and Garnet estimate that 40% of enterprise contact information will have leaked on Facebook by 2017. Unfortunately, interactions between legitimate enterprise applications and consumer applications are difficult to control.

The traditional employee badges will also be equipped with new sensors (apart from access control) thus tracking every move of the employee. This data will provide precious information about the employees which can help the organisation tackle issues before they actually arise. Imagine an employee who is excessively fidgety at his desk, it can be an indication that the employees' chair needs replacement or maybe the employee needs to control his dependency on caffeine. Intelligent clothing too will be introduced in the workplace. Cloths worn by employees will monitor their body temperature together with the ambience temperature and take action to tighten or loosen themselves in order to provide a constant comfortable temperature. This can help in optimising the use of cooling or heating in the office. It can also ensure that the employee is healthy and when the clothing detects an anomaly in the person's temperature, an alarm can be raised signalling the possibility of a fever. Future clothing will also allow us to charge our devices thus eliminating the problem of zero battery.

From a health point of view, humans will be better monitored and serviced. Their environment will alert them if some problems arise and will make a diagnosis. Personalised medicine will become the norm thus reducing risks and ensuring that the person undergoes a speedy recovery from his ailments. In case of bigger problems such as mobility challenges (especially in the case of disabled people), electronic exoskeletons[23] such as [108] will ensure that people still maintain their independence with the help of machines.

Robots will also make an appearance at the office. Their remit won't be restricted to cleaning but intelligent cameras will provide managers with valuable information about their employees. The robots will also have a social element thus allowing them to chat with human colleagues and exchange information with them. This will ensure that information (such as memos, etc) is shared in an effective way thus saving managers from having to do the rounds (especially in large dynamic organisations). Drones will also make an appearance in the office. Most probably they will be small and insect like with onboard cameras and indoor navigation. They will eventually replace the surveillance officers and in some cases, these drones will also allow real security officers (located at the control centre) to speak with employees.

The way employees interact with their machine will also change. Gone are the days of staring into a monitor. Spatial computing technology will make use of new surfaces such as doors or walls. It will also understand how we deal with physical

---

[23] External skeletons used to support and protect humans.

world objects, it will understand how we communicate and what is our relation to different places in the world. It will allow us to interact with a machine in a seamless way thus grabbing a file changes from being a search through folders in a hierarchical file system to grabbing (using gesture based technologies) a virtual representation of the file on a virtual desk.

9.7 Where Are We Now?

This was just a peep into what we might expect in the near future. Most of the technologies mentioned above are in a very advanced stage and we expect to see them commercialised in the coming decade. Notwithstanding this, we can never predict disruptive innovations which will change our world upside down. But remember, life is full of surprises so just be prepared and always hope for the best.

References

1. Usb: Device class definition for human interface devices (hid) (1996)
2. Rfid: The next generation of aidc. Technical report (2003)
3. United nations e-government survey 2012. Technical report (2012)
4. A Companion to Ancient Greek Government (Blackwell Companions to the Ancient World). Wiley-Blackwell (2013)
5. Social media can help prevent crimes: Delhi police (June 2013)
6. Adams, P.C.: Teaching and learning with simcity 2000. Journal of Geography 97(2), 47–55 (1998)
7. Allain, J.S.: From jeopardy! to jaundice: The medical liability implications of dr. watson and other artificial intelligence systems. La. L. Rev. 73, 1049–1183 (2013)
8. Amabile, T., Kramer, S.: The progress principle: Using small wins to ignite joy, engagement, and creativity at work. Harvard Business Press (2011)
9. Anil, A., Thampi, A.R., John, M.P., Shanthi, K.J.: Project haritha-an automated irrigation system for home gardens. In: 2012 Annual IEEE India Conference (INDICON), pp. 635–639. IEEE (2012)
10. Antle, A.: Supporting children's emotional expression and exploration in online environments. In: Proceedings of the 2004 Conference on Interaction Design and Children: Building a Community, pp. 97–104. ACM (2004)
11. Apperley, T.H.: Genre and game studies: Toward a critical approach to video game genres. Simulation & Gaming 37(1), 6–23 (2006)
12. Aron, J.: How innovative is apple's new voice assistant, siri? New Scientist 212(2836), 24 (2011)
13. Aron, J.: Smart kitchens keep novice chefs on track. New Scientist 215(2877), 17 (2012)
14. Ashruf, C.: Ambient intelligence, coming to you (May 2005)
15. Awschalom, D.D., Bassett, L.C., Dzurak, A.S., Hu, E.L., Petta, J.R.: Quantum spintronics: Engineering and manipulating atom-like spins in semiconductors. Science 339(6124), 1174–1179 (2013)
16. Axford, B., Browning, G.K., Huggins, R., Rosamond, B., Turner, J.: Politics: An Introduction. Routledge (2002)

17. Banyhamdan, K., Barakat, S.M.: Web 2.0: Internet technology used in human resource recruitment. American Academic & Scholarly Research Journal (AASRJ), 4(5) (2012)

18. Basso, M.: 2018: Digital natives grow up and rule the world. Gartner Research, 28 (2008)

19. Bejjanki, V.R., Zhang, R., Li, R., Pouget, A., Green, C.S., Lu, Z.-L., Bavelier, D.: Action video game play facilitates the development of better perceptual templates. Proceedings of the National Academy of Sciences (2014)

20. Kervin, L., Bennet, S., Maton, K.: The 'digital natives' debate: A critical review of the evidence. British Journal of Educational Technology (2008)

21. Benson, N., Collin, C., Ginsburg, J., Grand, V., Lazyn, M., Weeks, M.: The Psychology Book. Dorling Kindersley Publishers Ltd (2012)

22. Berners-Lee, T., Cailliau, R., Groff, J.-F.: The world-wide web. Computer Networks and ISDN Systems 25(4-5), 454–459 (1992)

23. Berners-Lee, T., Cailliau, R., Groff, J.-F.: The world-wide web. Computer Networks and ISDN Systems 25(4-5), 454–459 (1992)

24. Berners-Lee, T., Cailliau, R., Groff, J.-F.: The world-wide web. Computer Networks and ISDN Systems 25(4-5), 454–459 (1992)

25. Berners-Lee, T., Fischetti, M.: Weaving the Web: The Original Design and Ultimate Destiny of the World Wide Web by its Inventor. Harper San Francisco (September 1999)

26. Bhutta, C.B.: Not by the book facebook as a sampling frame. Sociological Methods & Research 41(1), 57–88 (2012)

27. Biemiller, A.: Vocabulary: Needed if more children are to read well. Reading Psychology 24(3-4), 323–335 (2003)

28. Binns, A.: Facebook's ugly sisters: Anonymity and abuse on formspring and ask. fm. Media Education Research Journal (2013)

29. Biocca, F., Levy, M.R.: Communication in the age of virtual reality. Routledge (1995)

30. Black, A.: Gen y: Who they are and how they learn. Educational Horizons 88(2), 92–101 (2010)

31. Blumenberg, H.: The genesis of the Copernican world. MIT Press (1987)

32. Boatright, R.G.: Campaign finance in the 2012 election. In: The American Elections of 2012, p. 145 (2013)

33. Boogers, M., Voerman, G.: Surfing citizens and floating voters: Results of an online survey of visitors to political web sites during the dutch 2002 general elections. Information Polity 8(1), 17–27 (2003)

34. Breuer, J., Festl, R., Quandt, T.: Digital war: An empirical analysis of narrative elements in military first-person shooters. Journal of Gaming & Virtual Worlds 4(3), 215–237 (2012)

35. Brown, B., Chalmers, M.: Tourism and mobile technology. In: ECSCW 2003, pp. 335–354. Springer (2003)

36. Brown, J., Bagci, I., King, A., Roedig, U.: Defend your home! jamming unsolicited messages in the smart home (2013)

37. Brown, N.J.: Egypt's failed transition. Journal of Democracy 24(4), 45–58 (2013)

38. Brown, S.: Wireless charging for all your devices ready for the market. Science (2014)

39. Browne, D.: Quantum computation: Model versus machine. Nature Physics 10(3), 179–180 (2014)

40. Brumbaugh, J.E.: Audel HVAC Fundamentals, Heating Systems, Furnaces and Boilers, Audel (2004)

41. Burgess, J., McLoughlin, C.: Investigating cyberbullying: Emerging research and e-safety strategies within families and communities. Communities, Children and Families Australia 6(1), 3 (2012)
42. Burkinshaw, R.: Alice and kev: The story of being homeless in the sims 3. Retrieved 19, 2010 (2009)
43. Bush, V.: As we think. The Atlantic Monthly 176(1), 101–108 (1945)
44. Buyya, R., Broberg, J., Goscinski, A.M.: Cloud Computing: Principles and Paradigms. Wiley Series on Parallel and Distributed Computing. Wiley (2011)
45. Byman, D.: Why drones work: The case for washington's weapon of choice (2013)
46. Cabrer, M.R., Redondo, R.P.D., Vilas, A.F., Arias, J.J.P., Duque, J.G.: Controlling the smart home from tv. IEEE Transactions on Consumer Electronics 52(2), 421–429 (2006)
47. Cazden, C., Cope, B., Fairclough, N., Gee, J.: A pedagogy of multiliteracies: Designing social futures. Harvard Educational Review 66(1), 60–92 (1996)
48. Chan, H., Fern, A., Ray, S., Wilson, N., Ventura, C.: Online planning for resource production in real-time strategy games. In: ICAPS, pp. 65–72 (2007)
49. Chapman, A.: Is sid meier's civilization history? Rethinking History 17(3), 312–332 (2013)
50. Chappell, L., Plas, J.V.: Technology drives advances in home health monitoring. WTN News (April 2008)
51. Chatfield, T.: Videogames now outperform hollywood movies. The Observer, 27 (2009)
52. ELIOT CHE. Securing a network society cyber-terrorism, international cooperation and transnational surveillance. Research Institute for European and American Studies (2007)
53. Chen, J.-H., Chi, P.-Y., Chu, H.-H., Chen, C.-H., Huang, P.: A smart kitchen for nutrition-aware cooking. IEEE Pervasive Computing 9(4), 58–65 (2010)
54. Chen, M.G.: Communication, coordination, and camaraderie in world of warcraft. Games and Culture 4(1), 47–73 (2009)
55. Chen, M.-H., Huang, L.-L., Lee, C.-F., Hsieh, C.-L., Lin, Y.-C., Liu, H., Chen, M.-I., Lu, W.-S.: A controlled pilot trial of two commercial video games for rehabilitation of arm function after stroke. Clinical rehabilitation, p. 0269215514554115 (2014)
56. Chen, S.-M., Wang, S., Lee, T.-Z., Lee, P.-H.: On the extent of importance and demand of the smart kitchen. Journal of Architecture & Planning 11(3), 243–273 (2010)
57. Chen, Y.-C., Chen, P., Song, R., Korba, L.: Online gaming crime and security issue-cases and countermeasures from taiwan (2004)
58. Chinn, M.D., Fairlie, R.W.: Ict use in the developing world: an analysis of differences in computer and internet penetration. Review of International Economics 18(1), 153–167 (2010)
59. Chomsky, N.: Knowledge of language: Its nature, origin, and use. Greenwood Publishing Group (1986)
60. Chui, M.: The social economy: Unlocking value and productivity through social technologies. McKinsey (2013)
61. Clayton, R.C., Murdoch, S.J., Watson, R.N.M.: Ignoring the great firewall of china. In: Danezis, G., Golle, P. (eds.) PET 2006. LNCS, vol. 4258, pp. 20–35. Springer, Heidelberg (2006)

62. Cleland, J.G.F., Swedberg, K., Follath, F., Komajda, M., Cohen-Solal, A., Aguilar, J.C., Dietz, R., Gavazzi, A., Hobbs, R., Korewicki, J., et al.: The euroheart failure survey programme – a survey on the quality of care among patients with heart failure in europe part 1: patient characteristics and diagnosis. European Heart Journal 24(5), 442–463 (2003)
63. Coffield, F., Moseley, D., Hall, E., Ecclestone, K., et al.: Learning styles and pedagogy in post-16 learning: A systematic and critical review (2004)
64. Benjamin, M.: Compaine. The digital divide: Facing a crisis or creating a myth? MIT Press (2001)
65. Cook, K.: Social recruiting: The role of social networking websites in the hiring practices of major advertising and public relations firms (2012)
66. Crawford, C.: The art of computer game design (1984)
67. Cruz-Neira, C., Sandin, D.J., DeFanti, T.A.: Surround-screen projection-based virtual reality: the design and implementation of the cave. In: Proceedings of the 20th Annual Conference on Computer Graphics and Interactive Techniques, pp. 135–142. ACM (1993)
68. Cutrí, G., Naccarato, G., Pantano, E.: Mobile cultural heritage: the case study of locri, pp. 410–420 (2008)
69. Dahlman, E., Parkvall, S., Skold, J., Beming, P.: 3G evolution: HSPA and LTE for mobile broadband. Access Online via Elsevier (2010)
70. Davies, R.J., Galway, L.B., Nugent, C.D., Jamison, C.H., Gawley, R.E., McCullagh, P.J., Zheng, H., Black, N.D.: A platform for self-management supported by assistive, rehabilitation and telecare technologies. In: 2011 5th International Conference on Pervasive Computing Technologies for Healthcare (PervasiveHealth), pp. 458–460. IEEE (2011)
71. de Macedo, D.V., Formico Rodrigues, M.A.: Experiences with rapid mobile game development using unity engine. Computers in Entertainment (CIE) 9(3), 14 (2011)
72. Deibert, R.: Black code: inside the battle for cyberspace. McClelland & Stewart, Toronto (2013)
73. Delahoussaye, M.: The perfect learner: An expert debate on learning styles. Training 39(5), 28–36 (2002)
74. Denning, T., Kohno, T., Levy, H.M.: Computer security and the modern home. Communications of the ACM 56(1), 94–103 (2013)
75. Derks, D., ten Brummelhuis, L.L., Zecic, D., Bakker, A.B.: Switching on and off..: Does smartphone use obstruct the possibility to engage in recovery activities? European Journal of Work and Organizational Psychology 23(1), 80–90 (2014)
76. Scott, P.D.: How iceland crowdsourced the creation of its new constitution (2011)
77. Dingli, A., Abela, C.: A pervasive assistant for nursing and doctoral staff. In: Proceedings of the Poster Track of the 18th European Conference on Artificial Intelligence (July 2008)
78. Dingli, A., Abela, C., D'Ambrogio, I.: Pervasive nursing and doctoral assistant — pinata. In: 2011 5th International Conference on Pervasive Computing Technologies for Healthcare (PervasiveHealth), pp. 123–130. IEEE (2011)
79. Dingli, A., Attard, D., Mamo, R.: Turning homes into low-cost ambient assisted living environments. International Journal of Ambient Computing and Intelligence (IJACI) 4(2), 1–23 (2012)
80. Dingli, A., Seychell, D.: Mobile edutainment in the city. mobile, p. 183 (2011)
81. Dingli, A., Seychell, D.: Mobile edutainment in the city. mobile, p. 183 (2011)

82. Dingli, A., Seychell, D.: Blending augmented reality with real world scenarios using mobile devices. Technologies and Protocols for the Future of Internet Design 258 (2012)

83. Dingli, A., Seychell, D.: Blending augmented reality with real world scenarios using mobile devices. Technologies and Protocols for the Future of Internet Design 258 (2012)

84. Dingli, A., Seychell, D.: Motivating learning through mobile interaction. In: Mobile Learning 2012, p. 271 (2012)

85. Dingli, A., Seychell, D.: Second generation digital natives. Technical report (2014)

86. Dingli, A., Wilks, Y., Catizone, R., Cheng, W.: The companions: Hybrid-world approach. In: 6th IJCAI Workshop on Knowledge and Reasoning in Practical Dialogue Systems, p. 60 (2009)

87. Dix, A., Finlay, J.E., Abowd, G.D., Beale, R.: Human-Computer Interaction, 3rd edn. Prentice-Hall, Inc., Upper Saddle River (2003)

88. Doherty, B.: John perry barlow 2.0 the thomas jefferson of cyberspace reinvents his body-and his politics. Reason-Santa Barbara Then Los Angeles, 36 (2004)

89. Donohue, L.: Nsa surveillance may be legal - but it's unconstitutional 6 (2013)

90. Drake, S.: Galileo Studies. The University of Michigan Press (1970)

91. Drexler, K.E., Minsky, M.: Engines of creation. Fourth Estate London (1990)

92. Druin, A.: The role of children in the design of new technology. Behaviour and Information Technology 21, 1–25 (2002)

93. Du Bois, E.C.: Woman Suffrage and Women's Rights. NYU Press (1998)

94. Dudley, A., Braman, J., Wang, Y., Vincenti, G., Tupper, D.: Security, legal, and ethical implications of using virtual worlds. In: Proceedings of the 14th World Multi-Conference on Systemics, Cybernetics and Informatics: WMSCI (2010)

95. Dwoskin, E.: Rand paul recruits for a class action against nsa (June 2013)

96. Earnshaw, R.A., Gigante, M.A., Jones, H.: Virtual reality systems, vol. 327. Academic Press, Orlando (1993)

97. Eichenwald, K.: On the web, pedophiles extend their reach. New York Times, 21 (2006)

98. Emanuel, E.J.: Online education: Moocs taken by educated few. Nature 503(7476), 342–342 (2013)

99. Essers, L.: Crowdsourced finnish copyright bill headed to parliament (July 2013)

100. Felder, R.M., Silverman, L.K.: Learning and teaching styles in engineering education. Engineering Education 78(7), 674–681 (1988)

101. Felder, R.M., Soloman, B.A.: Index of learning styles (1991)

102. Ferguson, C.J.: Does media violence predict societal violence? it depends on what you look at and when. Journal of Communication (2014)

103. Field, D., Catizone, R., Cheng, W., Dingli, A., Worgan, S., Ye, L., Wilks, Y.: The senior companion: a semantic web dialogue system. In: Proceedings of the 8th International Conference on Autonomous Agents and Multiagent Systems, vol. 2, pp. 1383–1384. International Foundation for Autonomous Agents and Multiagent Systems (2009)

104. Fleming, N., Baume, D.: Learning styles again: Varking up the right tree? Educational Developments 7(4), 4 (2006)

105. Flynn, J.R.: What is intelligence?: Beyond the Flynn effect. Cambridge University Press (2007)

106. Follaco, J.: Rit cybercrime research findings 6 (2008)

107. Forlizzi, J., Di Salvo, C.: Service robots in the domestic environment: a study of the roomba vacuum in the home. In: Proceedings of the 1st ACM SIGCHI/SIGART Conference on Human-Robot Interaction, pp. 258–265. ACM (2006)

108. French, J., Rollo, M., O'Malley, M.K.: [d75] advancements in robotic exoskeletons for upper limb rehabilitation. In: 2014 IEEE Haptics Symposium (HAPTICS), p. 1. IEEE (2014)

109. Freudenheim, M.: Wired up at home to monitor illnesses. New York Times, 23 (2010)

110. Fujinami, K., Kawsar, F., Nakajima, T.: AwareMirror: A personalized display using a mirror. In: Gellersen, H.-W., Want, R., Schmidt, A. (eds.) PERVASIVE 2005. LNCS, vol. 3468, pp. 315–332. Springer, Heidelberg (2005)

111. Gantz, J., Reinsel, D.: The digital universe in 2020: Big data, bigger digital shadows, and biggest growth in the far east. IDC iView: IDC Analyze the Future (2012)

112. Garcia-Ruiz, M.A., Martin, M.V.: A Ibrahim, A Edwards, and R Aquino-Santos. Combating child exploitation in second life. In: 2009 IEEE Toronto International Conference on Science and Technology for Humanity (TIC-STH), pp. 761–766. IEEE (2009)

113. Gaudiosi, J.: New reports forecast global video game industry will reach $82 billion by 2017. Forbes.com (2012)

114. Geist, M.A.: The legal implications of the yahoo! inc. nazi memorabilia dispute. Juriscom: January/March (2001)

115. CBRE Genesis. Fast forward, - the future of work and the workplace. Technical report (2030)

116. Gerbaudo, P.: Tweets and the Streets: Social Media and Contemporary Activism. Pluto Press (2012)

117. Giokas, K., Kouris, I., Koutsouris, D.: Autonomy, motivation and individual self-management for COPD patients, the amica project. In: Lin, J. (ed.) MobiHealth 2010. LNICST, vol. 55, pp. 49–53. Springer, Heidelberg (2011)

118. Gladwell, M.: Outliers: The story of success. Penguin UK (2008)

119. Gliozzo, A., Biran, O., Patwardhan, S., McKeown, K.: Semantic technologies in ibm watsontm. In: ACL 2013, p. 85 (2013)

120. Goldberg, D., Corcoran, M., Picard, R.G.: Remotely piloted aircraft systems & journalism: Opportunities and challenges of drones in news gathering. Reuters Institire for the Study of Journalism (June 2013)

121. Goldschlag, D., Levine, B.A.: Virtual reality: The reality of 2014 (2014)

122. Goodwin, S.: Smart Home Automation with Linux (Expert's Voice in Linux). Apress (2010)

123. Google and Ypsos. The New Multi-Screen World (2012)

124. Graser, M.: Videogames rock song sales (2009)

125. Greenberg, A.: This Machine Kills Secrets: How WikiLeakers, Cypherpunks, and Hacktivists Aim to Free the World's Information. Dutton Adult (2012)

126. Greenwald, G., MacAskill, E.: Nsa prism program taps in to user data of apple, google and others. The Guardian 7(6) (2013)

127. Gunkel, D.: Amazon drone delivery (2013)

128. Hagras, H., Callaghan, V., Colley, M., Clarke, G.: A hierarchical fuzzy–genetic multi-agent architecture for intelligent buildings online learning, adaptation and control. Information Sciences 150(1), 33–57 (2003)

129. Hai, J.C.: Fundamental of development administration (2007)

130. Hailpern, J., Guarino-Reid, L., Boardman, R., Annam, S.: Web 2.0: blind to an accessible new world. In: Proceedings of the 18th International Conference on World Wide Web, WWW 2009, pp. 821–830. ACM, New York (2009)

131. Hajinejad, N., Sheptykin, I., Grüter, B., Worpenberg, A., Lochwitz, A., Oswald, D., Vatterrott, H.-R.: Casual mobile gameplay–on integrated practices of research, design and play. In: Proceedings of the Think Design Play: The Fifth International Conference of the Digital Research Association (DIGRA), Hilversum, Netherlands (2011)

132. Hanak, D., Szijarto, G., Takacs, B.: A mobile approach to ambient assisted living. In: Proceedings of the IADIS Multi Conference on Computer Science and Information Systems 2007, pp. 3–8 (2007)

133. Harel, D., Feldman, Y.: Algorithmics: The spirit of computing. Pearson Education (2004)

134. Harjumaa, M., Isomursu, M., Muuraiskangas, S., Konttila, A.: Hearme: a touch-to-speech ui for medicine identification. In: 2011 5th International Conference on Pervasive Computing Technologies for Healthcare (PervasiveHealth), pp. 85–92. IEEE (2011)

135. Harper, R.: Inside the Smart Home. Springer (2003)

136. Hashimoto, A., Mori, N., Funatomi, T., Yamakata, Y., Kakusho, K., Minoh, M.: Smart kitchen: A user centric cooking support system. In: Proceedings of IPMU, vol. 8, pp. 848–854 (2008)

137. Hawking, S., Russell, S., Tegmark, M., Wilczek, F.: Transcendence looks at the implications of artificial intelligence-but are we taking ai seriously enough. The Independent (May 1, 2014)

138. Helal, S., Mann, W., El-Zabadani, H., King, J., Kaddoura, Y., Jansen, E.: The gator tech smart house: A programmable pervasive space. Computer 38(3), 50–60 (2005)

139. Helsper, E., Enyon, R.: Digital natives: Where is the evidence? British Education Research Journal (2009)

140. Hoffman, G., Moore, D.: Ieee 1394: A ubiquitous bus. In: Compcon 1995.'Technologies for the Information Superhighway', Digest of Papers, pp. 334–338. IEEE (1995)

141. Honey, P., Mumford, A.: The learning styles helper's guide. Peter Honey Maidenhead, Berkshire (2000)

142. Hong, X., Nugent, C.D., Devlin, S., Mulvenna, M.D., Wallace, J.G., Martin, S.: Assisted living services for reminding and prompting activities of daily living: A preliminary case study. In: PervasiveHealth, pp. 434–437 (2011)

143. Hooghe, M., Teepe, W.: Party profiles on the web: an analysis of the logfiles of non-partisan interactive political internet sites in the 2003 and 2004 election campaigns in belgium. New Media & Society 9(6), 965–985 (2007)

144. Horrigan, J.: Home broadband adoption 2009. Pew Internet & American Life Project (2009)

145. Hossain, A., Atrey, P.K., El Saddik, A.: Smart mirror for ambient home environment (2007)

146. Hourcade, J.P.: Interaction design and children. Found. Trends Hum.-Comput. Interact. 1(4), 277–392 (2008)

147. Hourcade, J.P.: Interaction design and children. Found. Trends Hum.-Comput. Interact. 1(4), 277–392 (2008)

148. Howard, P.N., Duffy, A., Freelon, D., Hussain, M., Mari, W., Mazaid, M.: Opening closed regimes: what was the role of social media during the arab spring? (2011)

149. Howe, N., Strauss, W.: Millennials rising: The next great generation. Vintage (2000)

150. Huang, C.: Facebook and twitter key to arab spring uprisings: report. The National. Abu Dhabi Media 6 (2011)

151. Huang, G.T.: Monitoring mom: As population matures, so do assisted-living technologies. Technical Review 20(1) (2003)
152. Huang, W., Webster, D.: Enabling context-aware agents to understand semantic resources on the wwwand the semantic web. In: Proceedings of the 2004 IEEE/WIC/ACM International Conference on Web Intelligence, WI 2004, pp. 138–144. IEEE Computer Society, Washington, DC (2004)
153. Huizinga, J., Francis, R.: Homo ludens. A study of the play-element in culture (Translated by RFC Hull). Routledge & Kegan Paul (1949)
154. Cisco Visual Networking Index. Forecast and methodology, 2011–2016, cisco systems. Inc., San Jose, CA (2012)
155. Inglis, S.: Structured telephone support or telemonitoring programmes for patients with chronic heart failure. Journal of Evidence-Based Medicine 3(4), 228–228 (2010)
156. Isken, M., Frenken, T., Brell, M., Hein, A.: Robot interaction with domestic environments considering aal services and smart home technologies. In: Intelligent Environments (Workshops), pp. 810–821 (2011)
157. Islam, S.S., Mollah, M.B., Huq, M.I., Aman Ullah, M.: Cloud computing for future generation of computing technology. In: 2012 IEEE International Conference on Cyber Technology in Automation, Control, and Intelligent Systems (CYBER), pp. 129–134 (2012)
158. Israel, R., Israel, R.C.: Global Citizenship-A Path to Building Identity and Community in a Globalized World. CreateSpace Independent Publishing Platform (2012)
159. Jaishankar, K.: Cyber criminology: exploring internet crimes and criminal behavior. CRC Press (2011)
160. Janczewski, L.J., Colarik, A.M.: Cyber warfare and cyber terrorism. IGI Global (2008)
161. Jara, A.J., Zamora, M.A., Skarmeta, A.F.: An architecture for ambient assisted living and health environments. In: Omatu, S., Rocha, M.P., Bravo, J., Fernández, F., Corchado, E., Bustillo, A., Corchado, J.M. (eds.) IWANN 2009, Part II. LNCS, vol. 5518, pp. 882–889. Springer, Heidelberg (2009)
162. Kanyuk, P., Young, J.: Rfid in healthcare: Novelty or mass market. Report BFTC1095 (2004)
163. Karlsson, M.K., Magnusson, H., von Schewelov, T., Rosengren, B.E.: Prevention of falls in the elderly–a review. Osteoporosis International 24(3), 747–762 (2013)
164. Kaufman, D., Sauvé, L.: Educational gameplay and simulation environments: Case studies and lessons learned. Information Science Reference (2010)
165. Kaur, I.: Microcontroller based home automation system with security. International Journal of Advanced Computer Science and Applications (IJACSA) 1(6), 60–65 (2010)
166. Kelly, B.B.: Investing in a centralized cybersecurity infrastructure: Why "hacktivism" can and should influence cybersecurity reform. Bul. Rev. 92, 1663–1663 (2012)
167. Khondker, H.H.: Role of the new media in the arab spring. Globalizations 8(5), 675–679 (2011)
168. Kissane, D.: Chasing the youth vote: Kevin07, web 2.0 and the 2007 australian federal election (2008)
169. Kissane, D.: Kevin07, web 2.0 and young voters at the 2007 australian federal election. CEU Political Science Journal (02), 144–168 (2009)
170. Kolb, A.Y.: The kolb learning style inventory–version 3, vol. 200. Hay Resource Direct, Boston (2005)
171. Kolb, D.A., et al.: Experiential learning: Experience as the source of learning and development, vol. 1. Prentice-Hall, Englewood Cliffs (1984)

172. Komajda, M., Follath, F., Swedberg, K., Cleland, J., Aguilar, J.C., Cohen-Solal, A., Dietz, R., Gavazzi, A., Van Gilst, W.H., Hobbs, R., et al.: The euroheart failure survey programme-a survey on the quality of care among patients with heart failure in europe part 2: treatment. European Heart Journal 24(5), 464–474 (2003)
173. Koyre, A.: From the Closed World to the Infinite Universe. Johns Hopkins University Press (1957)
174. Kuhn, T.: The Copernican Revolution. Planetary Astronomy in the Development of Western Thought. Harvard University Press (1957)
175. Kuhn, T.: Logic of Discovery or Psychology of Research. Cambridge University Press (1972)
176. Kuhn, T.S.: The structure of scientific revolutions. University of Chicago Press, Chicago (1970)
177. Lah, K.: Tokyo man marries video game character. CNN.com, 17 (2009)
178. LaMonica, M.: Nest thermostat slays peak power (April 2013)
179. Lange, B., Flynn, S., Rizzo, A.: Initial usability assessment of off-the-shelf video game consoles for clinical game-based motor rehabilitation. Physical Therapy Reviews 14(5), 355–363 (2009)
180. Lanting, T., Przybysz, A.J., Yu Smirnov, A., Spedalieri, F.M., Amin, M.H., Berkley, A.J., Harris, R., Altomare, F., Boixo, S., Bunyk, P., et al.: Entanglement in a quantum annealing processor. arXiv preprint arXiv:1401.3500 (2014)
181. Leach, G.J., Sugarman, T.S.: Play to win! using games in library instruction to enhance student learning. Research Strategies 20(3), 191–203 (2005)
182. Lee, H., Park, S.J., Kim, M.J., Lim, H.W., Kim, J.T.: The energy aware system of a smart bedroom for single person homes. In: 7th International Symposium on Sustainable Healthy Buildings, Seoul, Korea (2012)
183. Lee, J.J., Hammer, J.: Gamification in education: What, how, why bother? Academic Exchange Quarterly 15(2), 146 (2011)
184. Lee, N.: Cyber attacks, prevention, and countermeasures. In: Counterterrorism and Cybersecurity, pp. 119–142. Springer (2013)
185. Lehdonvirta, V.: Virtual item sales as a revenue model: identifying attributes that drive purchase decisions. Electronic Commerce Research 9(1-2), 97–113 (2009)
186. Lenhart, A., Kahne, J., Middaugh, E., Macgill, A.R., Evans, C., Vitak, J.: Teens, video games, and civics: Teens' gaming experiences are diverse and include significant social interaction and civic engagement. Pew Internet & American Life Project (2008)
187. Minister Sabine Leutheusser-Schnarrenberger. Us prism scandal: šecurity is not an end in itself (2013)
188. Lian, Z., Mei, Z., et al.: Application of automatic water-saving irrigation system in roof gardens. Journal of Landscape Research 1(4), 75–79 (2009)
189. Liao, W.-H., Kuo, J.-H., Yang, C.-M., Chen, I.Y.: iwakeup: A video-based alarm clock for smart bedrooms. Journal of the Chinese Institute of Engineers 33(5), 661–668 (2010)
190. Liles, S.: Cyber warfare: As a form of low–intensity conflict and insurgency. In: Conference on Cyber Conflict Proceedings, pp. 47–58 (2010)
191. Limber, S.P.P., Agatston, P.W.W.: Cyberbullying. Blackwell Publishing (2012)
192. Linehan, C., Waddington, J., Hodgson, T.L., Hicks, K., Banks, R.: Designing games for the rehabilitation of functional vision for children with cerebral visual impairment. In: CHI 2014 Extended Abstracts on Human Factors in Computing Systems, pp. 1207–1212. ACM (2014)
193. Lipson, H., Kurman, M.: Fabricated: The new world of 3D printing. John Wiley & Sons (2013)

194. Lobosco, K.: Army of robots to invade amazon warehouses. CNN Money (May 2014)
195. Lombardi, A., Ferri, M., Rescio, G., Grassi, M., Malcovati, P.: Wearable wireless accelerometer with embedded fall-detection logic for multi-sensor ambient assisted living applications. In: 2009 IEEE Sensors, pp. 1967–1970. IEEE (2009)
196. MacFarlane, S., Pasiali, A.: Adapting the Heuristic Evaluation Method for Use with Children. In: Workshop on Child Computer Interaction: Methodological Research at Interact, Rome, Italy (2005)
197. Macnamara, J., Kenning, G.: E-electioneering 2010: Trends in social media use in australian political communication. Media International Australia, Incorporating Culture & Policy (139), 7 (2011)
198. Magjarević, R.: Health care issues within ambient assisted living. In: ISBME 2006-2nd International Symposium on Biomedical Engineering (2006)
199. Mainemelis, C., Boyatzis, R.E., Kolb, D.A.: Learning styles and adaptive flexibility testing experiential learning theory. Management Learning 33(1), 5–33 (2002)
200. Mani, A., Rahwan, I., Pentland, A.: Inducing peer pressure to promote cooperation. Scientific Reports 3 (2013)
201. Mann, S.: Intelligent bathroom fixtures and systems: Existech corporation's safebath project. Leonardo 36(3), 207–210 (2003)
202. Mannoor, M.S., Jiang, Z., James, T., Kong, Y.L., Malatesta, K.A., Soboyejo, W.O., Verma, N., Gracias, D.H., McAlpine, M.C.: 3d printed bionic ears. Nano Letters 13(6), 2634–2639 (2013)
203. Marriott, T.: Designing social behaviour through play (2011)
204. Martinet, P., Laugier, C., Nunes, U.: Special issue on perception and navigation for autonomous vehicles. IEEE Robotics & Automation Magazine (2014)
205. Marwick, A.E., Diaz, D.M., Palfrey, J.: Youth, privacy and reputation. Literature Review. The Berkman Center for Internet & Society at Harvard University 9:2010 (2010), http://cyber.law.harvard.edu/publications/2010/Youth_Privacy_Reputation_Lit_Review
206. Mattern, F.: Vom verschwinden des computers–die vision des ubiquitous computing. In: Total Vernetzt, pp. 1–41. Springer (2003)
207. Matthews, D.: The rise and fall of the anti-counterfeiting trade agreement (acta): Lessons for the european union. Queen Mary School of Law Legal Studies Research Paper (127) (2012)
208. Mazuryk, T., Gervautz, M.: Virtual reality-history, applications, technology and future (1996)
209. Mcardle, M.: The rise of the artifical-intelligence economy. The Atlantic (April 2012)
210. McCarra, D.: Wikipedia community has spoken, global anti-sopa blackout protest begins wednesday (2012)
211. McCaulley, M.H.: Jung's theory of psychological types and the Myers-Briggs Type Indicator. Center for applications of Psychological Type (1981)
212. McClarty, K.L., Orr, A., Frey, P.M., Dolan, R.P., Vassileva, V., McVay, A.: A literature review of gaming in education. Gaming in Education (2012)
213. McDevitt, T.M., Ormrod, J.E.: Child development and education. Pearson College Div. (2009)
214. McGonigal, J.: Reality is broken: Why games make us better and how they can change the world. Penguin (2011)
215. McGregor, L.: George Orwell's Animal farm. Barron's, New York (1999)
216. Meng, X., Qiang, L., Wei, J., Shi, H.: Preparation of electrophoretic nanoparticles for electronic paper. Journal of Nanoscience and Nanotechnology 14(2), 1617–1630 (2014)

217. Metaxas, P.T., Mustafaraj, E., Gayo-Avello, D.: How (not) to predict elections. In: 2011 IEEE Third International Conference on Privacy, Security, Risk and Trust (PASSAT) and 2011 IEEE Third International Conference on Social Computing (SocialCom), pp. 165–171. IEEE (2011)

218. Meyer, G.: Smart Home Hacks: Tips & Tools for Automating Your House. O'Reilly Media (2004)

219. Meyer, J., Boll, S.: Digital health devices for everyone! IEEE Pervasive Computing 13(2), 10–13 (2014)

220. Milgram, P., Takemura, H., Utsumi, A., Kishino, F.: Augmented reality: A class of displays on the reality-virtuality continuum. In: Photonics for Industrial Applications, pp. 282–292. International Society for Optics and Photonics (1995)

221. Mironov, V., Boland, T., Trusk, T., Forgacs, G., Markwald, R.R.: Organ printing: computer-aided jet-based 3d tissue engineering. Trends in Biotechnology 21(4), 157–161 (2003)

222. Mischuck, B., Mølmer, K.: Qudit quantum computation in the jaynes-cummings model. Physical Review A 87(2), 022341 (2013)

223. Mitchell, A., Hitlin, P.: Twitter reaction to events often at odds with overall public opinion. Technical report (March 2013)

224. Monegain, B.: Home monitoring gives heart patients a boost. Healthcare IT News (June 2010)

225. Moore, G.E.: Cramming more components onto integrated circuits. Electronics 38(8) (April 1965)

226. Morato, J., Fraga, A., Andreadakis, Y., Sánchez-Cuadrado, S.: Semantic web or web 2.0? socialization of the semantic web. In: Lytras, M.D., Carroll, J.M., Damiani, E., Tennyson, R.D., Avison, D., Vossen, G., De Pablos, P.O. (eds.) WSKS 2008. CCIS, vol. 19, pp. 406–415. Springer, Heidelberg (2008)

227. Myers, D.: The nature of computer games. Peter Lang, New York (2003)

228. Nielsen, J.: Heuristic Evaluation. In: Usability Inspection Methods, pp. 25–62. John Wiley & Sons, Inc., New York (1994)

229. Dept of Physics and Iowa State University Astronomy. The ptolematic model. Online (2001)

230. Oh, H., Bahn, H., Chae, K.-J.: An energy-efficient sensor routing scheme for home automation networks. IEEE Transactions on Consumer Electronics 51(3), 836–839 (2005)

231. Olson, T., Nelsonl, T.: The internet's impact on political parties and campaigns. International Reports (May 2010)

232. O'Reilly, T.: What is web 2.0? In: O'Reilly Media Conference (2005)

233. O'reilly, T.: What is web 2.0. O'Reilly Media, Inc. (2009)

234. Osborn, A., Young, S.: Uk government rejects acusations its use of u.s. spy system was illegal (June 2013)

235. Amutha, K.P., Sethukkarasi, C., Pitchiah, R.: Smart kitchen cabinet for aware home. In: The First International Conference on Smart Systems, Devices and Technologies, SMART 2012, pp. 9–14 (2012)

236. Palfrey, J., Gasser, U.: Born Digital: Understanding the First Generation of Digital Natives. Basic Books, Inc., New York (2008)

237. Palfrey, J.G., Gasser, U.: Born digital: Understanding the first generation of digital natives. Basic Books (2013)

238. Park, S.H., Won, S.H., Lee, J.B., Kim, S.W.: Smart home: digitally engineered domestic life. Personal and Ubiquitous Computing 7(3-4), 189–196 (2003)

239. Pashler, H., McDaniel, M., Rohrer, D., Bjork, R.: Learning styles concepts and evidence. Psychological Science in the Public Interest 9(3), 105–119 (2008)

240. Patterson, D., Hennessy, J.: Computer Organisation and Design: The Hardware-Software Interface (2004)
241. Patterson, D., Hennessy, J.: Computer Organisation and Design: The Hardware-Software Interface (2004)
242. Pease, P.S.: The virtual university: Jones international university, ltd. Information, Communication & Society 3(4), 627–628 (2000)
243. Peng, H., Huang, Z.C.: Design of a type of cleaning robot system. Advanced Materials Research 605, 1415–1418 (2013)
244. Pepitone, J.: Sopa and pipa postponed indefinitely after protests. CNN Money 20 (2012)
245. Mell, P., Grance, T.: The NIST Definition of Cloud Computing (2009)
246. Pettey, C., van der Meulen, R.: Gartner says free apps will account for nearly 90 percent of total mobile app store downloads in 2012. Gartner (September 11, 2012)
247. Piaget, J.: Part i: Cognitive development in children: Piaget development and learning. Journal of Research in Science Teaching 2(3), 176–186 (1964)
248. Piliouras, T., Yu, R., Villanueva, K., Chen, Y., Robillard, H., Berson, M., Lauer, J., Sampel, G., Lapinski, D., Attre, M.: A deeper understanding of technology is needed for workforce readiness-playing games, texting, and tweets aren't enough to make students tech-savvy. In: 2014 Zone 1 Conference of the American Society for Engineering Education (ASEE Zone 1), pp. 1–8. IEEE (2014)
249. Pinto, H., Wilks, Y., Catizone, R., Dingli, A.: The senior companion multiagent dialogue system. In: Proceedings of the 7th International Joint Conference on Autonomous Agents and Multiagent Systems, vol. 3, pp. 1245–1248. International Foundation for Autonomous Agents and Multiagent Systems (2008)
250. Pistono, F.: Robots will steal your job, but that's ok: how to survive the economic collapse and be happy. Federico Pistono (2012)
251. Plato: The Republic. Penguin Classics (2012)
252. Plouffe, D.: The audacity to win: the inside story and lessons of Barack Obama's historic victory. Viking, New York (2009)
253. Pogorelc, B.: An ubiquitous and intelligent system for prolonging independent living of elderly users. IEEE (April 2012)
254. Prensky, M.: Digital natives, digital immigrants (2001)
255. Prensky, M.: Digital natives, digital immigrants part 1. On the Horizon 9(5), 1–6 (2001)
256. Prensky, M.: Digital Natives, Digital Immigrants, Part II: Do They Really Think Differently? On the Horizon (2001)
257. Oxford University Press. Definition of smart home (December 2012),
      https://oxforddictionaries.com/
      definition/english/smart+home
258. Pressman, R.: Software Engineering: A Practitioner's Approach, 6th edn. McGraw-Hill, Inc., New York (2005)
259. Prichard, J.J., MacDonald, L.E.: Cyber terrorism: A study of the extent of coverage in computer security textbooks. Journal of Information Technology Education 3, 279–289 (2004)
260. Prokhorov, S.: Social media and democracy: Facebook as a tool for the establishment of democracy in egypt (2012)
261. Qualman, E.: Socialnomics: How social media transforms the way we live and do business. John Wiley & Sons (2012)
262. Read, J., Gregory, P., Macfarlane, S., Mcmanus, B., Gray, P., Patel, R.: An investigation of participatory design with Children – Informant, Balanced and Facilitated Design. In: Proceedings of Interaction Design and Children International Workshop, pp. 53–64. Shaker Publishing (2002)

263. ABI Research. 90 million homes worldwide will employ home automation systems by 2017 (May 2012)
264. Reynolds, D.: Structured home monitoring system improves outcomes for heart failure patients. Emax Health (August 2010)
265. Rice, R.E., Katz, J.E.: Comparing internet and mobile phone usage: digital divides of usage, adoption, and dropouts. Telecommunications Policy 27(8), 597–623 (2003)
266. Rieffel, E.: Quantum computing: a gentle introduction. MIT Press, Cambridge (2011)
267. Roberts, P.: The Impulse Society: America in the Age of Instant Gratification. Bloomsbury Publishing, USA (2014)
268. Robles, R.J., Kim, T.-H.: A review on security in smart home development. International Journal of Advanced Science and Technology 15 (2010)
269. Rogers, Y., Sharp, H., Preece, J.: Interaction design: Beyond human-computer interaction (2011)
270. Rollings, A., Morris, D.: Game architecture and design: a new edition (2003)
271. Rosin, H.: The touch-screen generation. The Atlantic (2013)
272. Rosser, J.C., Lynch, P.J., Cuddihy, L., Gentile, D.A., Klonsky, J., Merrell, R.: The impact of video games on training surgeons in the 21st century. Archives of Surgery 142(2), 181–186 (2007)
273. Roth, I.: Die internationale Zuständigkeit deutscher Gerichte bei Persönlichkeitsrechtsverletzungen im Internet. Lang (2007)
274. Rubel, P., Fayn, J., Simon-Chautemps, L., Atoui, H., Ohlsson, M., Telisson, D., Adami, S., Arod, S., Forlini, M.C., Malossi, C., et al.: New paradigms in telemedicine: ambient intelligence, wearable, pervasive and personalized. Stud. Health Technol. Inform. 108, 123–132 (2004)
275. Russell, G.: Pedophiles in wonderland: censoring the sinful in cyberspace. J. Crim. L. & Criminology 98, 1467 (2007)
276. Salen, K.: Quest to learn: Developing the school for digital kids. MIT Press (2011)
277. Saletan, W.: Springtime for twitter: Is the internet driving the revolutions of the arab spring. Slate (2011), http://www.slate.com/id/2299214/
278. Savitz, E.: Gartner: Top 10 strategic technology trends for 2013 (October 2012)
279. Schiesel, S.: Social significance in playing online? you betcha! New York Times D 7 (2005)
280. Schneider, M.: The semantic cookbook: sharing cooking experiences in the smart kitchen. In: 3rd IET International Conference on Intelligent Environments, IE 2007, pp. 416–423. IET (2007)
281. Von Schweber, L., Von Schweber, E.: Virtual reality: Virtually here. PC Magazine-Boulder 14(5), 168–200 (1995)
282. Schweitzer, E.J.: Innovation or normalization in e-campaigning? a longitudinal content and structural analysis of german party websites in the 2002 and 2005 national elections. European Journal of Communication 23(4), 449–470 (2008)
283. Senoh, T., Ichihashi, Y., Oi, R., Sasaki, H., Yamamoto, K.: Study of a holographic tv system based on multi-view images and depth maps. In: SPIE OPTO, pp. 86440A–86440A. International Society for Optics and Photonics (2013)
284. Seo, Y.-H.: High-performance digital holographic video system. In: Digital Holography and Three-Dimensional Imaging, pp. DW2A–10. Optical Society of America (2013)
285. Seybert, H.: Internet use in households and by individuals in 2011. Eurostat Statistics in Focus 66 (2011)
286. Sheldon, F.T., McDonald, T.: Introduction to the special issue on cyber security and management. Information Systems and e-Business Management, 1–3 (2012)

287. Shih, P.-S., Cheng, J.-S.: Electronic paper display device (January 2014) US Patent 8,634,12

288. Shudong, W., Higgins, M.: Limitations of mobile phone learning. In: IEEE International Workshop on Wireless and Mobile Technologies in Education, WMTE 2005, p. 3. IEEE (2005)

289. Silberstein, E.: Residential Construction Academy HVAC. Cengage Learning (2011)

290. Singhal, A.: Modern Information Retrieval: A Brief Overview. Bulletin of the IEEE Computer Society Technical Committee on Data Engineering 24(4), 35–42 (2001)

291. Singla, N.: Wireless charging of mobile phone using microwaves or radio frequency signals (2014)

292. Skinner, B.F.: The technology of teaching (1968)

293. Slater, M., Wilbur, S.: A framework for immersive virtual environments (five): Speculations on the role of presence in virtual environments. Presence: Teleoperators and Virtual Environments 6(6), 603–616 (1997)

294. Smet, H., Smet, J., Cuypers, D., Weng, C.-P., Joshi, P.: Paper no 3.1: A contact lens with built-in display: Science fiction or not? In: SID Symposium Digest of Technical Papers, vol. 44, pp. 8–11. Wiley Online Library (2013)

295. Sottek, T.C.: The declaration of internet freedom: how the net's minutemen plan to protect the future (2012)

296. Sriskanthan, N., Tan, F., Karande, A.: Bluetooth based home automation system. Microprocessors and Microsystems 26(6), 281–289 (2002)

297. Stachowiak, M.E.: Automated dispensing cabinets: curse or cure? AJN The American Journal of Nursing 113(5), 11 (2013)

298. Stallings, W.: Computer Organization and Architecture, 7th edn. Prentice-Hall, Inc., Upper Saddle River (2006)

299. Stander, M., Hadjakos, A., Lochschmidt, N., Klos, C., Renner, B., Muhlhauser, M.: A smart kitchen infrastructure. In: 2012 IEEE International Symposium on Multimedia (ISM), pp. 96–99. IEEE (2012)

300. Stanyer, J.: Online campaign communication and the phenomenon of blogging: an analysis of web logs during the 2005 british general election campaign. In: Aslib Proceedings, vol. 58, pp. 404–415. Emerald Group Publishing Limited (2006)

301. EPA's Energy Star. A Guide to Energy-Efficient Heating and Cooling. CreateSpace Independent Publishing Platform (2013)

302. Steffen, D., Bleser, G., Weber, M., Stricker, D., Fradet, L., Marin, F.: A personalized exercise trainer for elderly. In: 2011 5th International Conference on Pervasive Computing Technologies for Healthcare (PervasiveHealth), pp. 24–31. IEEE (2011)

303. Stein, B., Lipsher, D.: The Value of Human Capital in the Digital Age. Korn/Ferry Institute (2013)

304. Stepanova, E.: The role of information communication technologies in the 'arab spring'. Ponars Eurasia (15), 1–6 (2011)

305. Stice, J.E.: Using kolb's learning cycle to improve student learning. Engineering Education 77(5), 291–296 (1987)

306. Tan, M., Su, X.: Media cloud: When media revolution meets rise of cloud computing. In: 2011 IEEE 6th International Symposium on Service Oriented System Engineering (SOSE), pp. 251–261. IEEE (2011)

307. Tapscott, D.: Growing up digital: The rise of the net generation (1998)

308. Thikey, H., van Wjick, F., Grealy, M., Rowe, P.: A need for meaningful visual feedback of lower extremity function after stroke. In: 2011 5th International Conference on Pervasive Computing Technologies for Healthcare (PervasiveHealth), pp. 379–383. IEEE (2011)

309. Thomas, M.: Deconstructing Digital Natives: Young people, technology, and the new literacies. Taylor & Francis (2011)

310. Toor, A.: Nutrismart prototype embeds rfid tags directly within food, traces your lunch from start to finish (May 2011)

311. Ueno, T., Inada, R., Saeki, O., Tsuji, K.: Effectiveness of displaying energy consumption data in residential houses. Proceedings of the European Council for an Energy-Efficient Economy 6, 19 (2005)

312. Vaccari, C.: Research note: Italian parties' websites in the 2006 elections. European Journal of Communication 23(1), 69–77 (2008)

313. Vaccari, C.: Surfing to the elysee: The internet in the 2007 french elections. French Politics 6(1), 1–22 (2008)

314. Vaccari, C.: From echo chamber to persuasive device? rethinking the role of the internet in campaigns. New Media & Society 15(1), 109–127 (2013)

315. Van Den Broek, G., Cavallo, F., Wehrman, C.: Aaliance Ambient Assisted Living Roadmap, vol. 6. IOS Press (2010)

316. van Wamelen, J., de Kool, D.: Web 2.0: a basis for the second society? In. In: Proceedings of the 2nd International Conference on Theory and Practice of Electronic Governance, ICEGOV 2008, pp. 349–354. ACM, New York (2008)

317. Verny, T., Kelly, J.: The Secret Life of the Unborn Child: How You Can Prepare Your Baby for a Happy, Healthy Life. Dell (1982)

318. Vygotsky, L.: Interaction between learning and development. Readings on the Development of Children, pp. 34–41 (1978)

319. Waite, M.: Journalism with flying robots. XRDS: Crossroads, The ACM Magazine for Students 20(3), 28–31 (2014)

320. Wakabayashi, D.: Mirrors that double as computers: High-tech looking glasses can display vital signs, or aid in rehabilitation. The Wall Street Journal (September 2012)

321. Wang, Y.D., Zahadat, N.: Teaching web development in the web 2.0 era. In: Proceedings of the 10th ACM Conference on SIG-Information Technology Education, SIGITE 2009, pp. 80–86. ACM, New York (2009)

322. Watson, J.: Rise of the machines: Moving from hype to reality in the burgeoning market for machine-to-machine communication (2012)

323. Wattal, S., Schuff, D., Mandviwalla, M., Williams, C.B.: Web 2.0 and politics: the 2008 us presidential election and an e-politics research agenda. Mis Quarterly 34(4), 669–688 (2010)

324. Webb, W.: The HTTP Murders: 15 Cyber Killers You Never Want to Meet Online. Absolute Crime (2013)

325. Weitzenboeck, E.M.:... still we are left wanting: Malta's white paper on digital rights. Computer Law & Security Review 29(3), 293–295 (2013)

326. Werner, F., Diermaier, J., Schmid, S., Panek, P.: Fall detection with distributed floor-mounted accelerometers: An overview of the development and evaluation of a fall detection system within the project ehome. In: 2011 5th International Conference on Pervasive Computing Technologies for Healthcare (PervasiveHealth), pp. 354–361. IEEE (2011)

327. Wilks, Y., Catizone, R., Worgan, S., Dingli, A., Moore, R., Field, D., Cheng, W.: A prototype for a conversational companion for reminiscing about images. Computer Speech & Language 25(2), 140–157 (2011)

328. Will, C.M.: "relativity". grolier multimedia encyclopaedia. Online (2010)

329. Will, S.: Uk workers deal with 10,000 emails a year, new research discovers. Technical report (April 2013)

330. Wilson, C.: Botnets, cybercrime, and cyberterrorism: Vulnerabilities and policy issues for congress. DTIC Document (2008)
331. Wilson, M.W.: The new role of the small donor in political campaigns and the demise of public funding. JL & Pol. 25, 257 (2009)
332. Windisch, E., Medman, N.: Understanding the digital natives. Ericsson Business Review - 1/2008 (2008)
333. Witmer, B.G., Singer, M.J.: Measuring presence in virtual environments: A presence questionnaire. Presence: Teleoperators and Virtual Environments 7(3), 225–240 (1998)
334. Wolak, J., Finkelhor, D., Mitchell, K.J., Ybarra, M.L.: Online "predators" and their victims. Psychology of Violence 1, 13–35 (2010)
335. Yakovlev, I.V.: Web 2.0: Is it evolutionary or revolutionary? IT Professional 9(6), 43–45 (2007)
336. Young, S.: Uk's cameron defends spy agencies over prism cyber-snooping (June 2013)
337. Zhang, S., Zhang, S., Chen, X., Huo, X.: Cloud computing research and development trend. In: Second International Conference on Future Networks, ICFN 2010, pp. 93–97 (2010)
338. Zhang, S., Zhang, S., Chen, X., Huo, X.: Cloud computing research and development trend. In: Second International Conference on Future Networks, ICFN 2010, pp. 93–97 (January 2010)
339. Zheng, P., Ni, L.M.: Spotlight: the rise of the smart phone. IEEE Distributed Systems Online 7(3) (2006)
340. Zhou, Y., Guan, X., Zheng, Q., Sun, Q., Zhao, J.: Group dynamics in discussing incidental topics over online social networks. IEEE Network 24(6), 42–47 (2010)
341. Zichermann, G., Cunningham, C.: Gamification by design: Implementing game mechanics in web and mobile apps. O'Reilly Media, Inc. (2011)
342. Zimmer, C.: How google is making us smarter. Discover Magazine 30(2), 30–32 (2009)
343. Zur, O., Zur, A.: On digital immigrants and digital natives: How the digital divide affects families, educational institutions, and the workplace. Zur Institute (2011) (retrieved)

Index

Printed in the United States
By Bookmasters